Econometric Analyses of International Housing Markets

This book explores how econometric modelling can be used to provide valuable insight into international housing markets. Initially describing the role of econometrics modelling in real estate market research and how it has developed in recent years, the book goes on to compare and contrast the impact of various macroeconomic factors on developed and developing housing markets.

Explaining the similarities and differences in the impact of financial crises on housing markets around the world, the author's econometric analysis of housing markets across the world provides a broad and nuanced perspective on the impact of both international financial markets and local macro economy on housing markets. With discussion of countries such as China, Germany, UK, US and South Africa, the lessons learned will be of interest to scholars of real estate economics around the world.

Rita Yi Man Li received her undergraduate and postgraduate degrees from the University of Hong Kong. She now serves as an Associate Professor in the Department of Economics and Finance, Hong Kong Shue Yan University (HKSYU). She is also a founder and Director of the Sustainable Real Estate Research Center and an Adjunct Associate Professor in the China-Australia Centre for Sustainable Urban Development, Tianjin University. She first started her academic career as a Visiting Lecturer at Hong Kong Polytechnic University, then a Lecturer, an Assistant Professor and now an Associate Professor at HKSYU. Outside the academia, she is a chartered surveyor by profession, an associate member of CIArb and HKIArb.

Over the years, she has won many local and international awards and grants. She is the recipient of the RGC grant in Hong Kong, Ronald Coase Institute Fellowship in the US and a competitive fellowship from the Australian government. She is a journal editor/editor-in-chief for many international academic peer-reviewed journals and a regular reviewer for top journals in the built environment. Her research activities have been covered by mass media in Hong Kong, Macau, the UK, France and mainland China.

Kwong Wing Chau is currently Head and Chair Professor of Real Estate and Construction and Director of the Ronald Coase Centre for Property Rights Research at the University of Hong Kong. His main areas of research include real estate finance and economics, real estate price index and institutional analysis of the built environment. Most of his works are empirical studies with implications for policy makers and practitioners. He has published over 100 journal papers including in the Journal of Law and Economics, Journal of Real Estate Finance and Economics, Land Use Policy, Urban Studies, Energy Policy, Building and Environment, Environment and Planning B and Habitat International. He received the International Real Estate Society Achievement Award in 1999. In 1996, Dr Chau became the founding president of the Asian Real Estate Society. He was elected president of the International Real Estate Society in 2000. He is now serving on the editorial boards of more than 10 peer-reviewed journals.

Routledge Studies in International Real Estate

The Routledge Studies in International Real Estate series presents a forum for the presentation of academic research into international real estate issues. Books in the series are broad in their conceptual scope and reflect an inter-disciplinary approach to Real Estate as an academic discipline.

Oiling the Urban Economy
Land, labour, capital, and the state in Sekondi-Takoradi, Ghana
Franklin Obeng-Odoom

**Real Estate, Construction and Economic Development
in Emerging Market Economies**
*Edited by Raymond T. Abdulai, Franklin Obeng-Odoom,
Edward Ochieng and Vida Maliene*

Econometric Analyses of International Housing Markets
Rita Li and Kwong Wing Chau

Econometric Analyses of International Housing Markets

Rita Yi Man Li and
Kwong Wing Chau

Routledge
Taylor & Francis Group

LONDON AND NEW YORK

First published 2016 by Routledge

2 Park Square, Milton Park, Abingdon, Oxon, OX14 4RN
605 Third Avenue, New York, NY 10017

Routledge is an imprint of the Taylor & Francis Group, an informa business

First issued in paperback 2020

British Library Cataloguing-in-Publication Data
A catalogue record for this book is available from the British Library

Library of Congress Cataloging-in-Publication Data
Names: Li, Rita Yi Man, author. | Chau, K.W. (Kwong Wing) author.
Title: Econometric analyses of international housing markets / Rita Yi Man Li and K.W Chau.
Description: New York : Routledge, 2016. | Series: Routledge studies in international real estate | Includes bibliographical references and index.
Identifiers: LCCN 2015042818 | ISBN 9781138821934
    (hardback : alk. paper) | ISBN 9781315743035 (ebook)
Subjects: LCSH: Housing—Prices—Econometric models. | Housing policy.
Classification: LCC HD7287.L5 2016 | DDC 333.33/8015195—dc23
LC record available at http://lccn.loc.gov/2015042818

ISBN: 978-1-138-82193-4 (hbk)
ISBN: 978-0-367-73719-1 (pbk)

Typeset in Times
by Swales & Willis Ltd, Exeter, Devon, UK

Contents

1 Introduction

Rita Yi Man Li and Kwong Wing Chau

Housing not only provides shelter for human habitation, it also plays an important role in the macroeconomy. As home purchase often needs the input of workmen from various sectors, homebuyers not only pay for the housing unit alone but also provide jobs and business opportunities for various sectors. For example, lawyers provide legal advice and services, bankers provide mortgage services, while construction workers and designers help decorate houses. Hence, the flourishing and withering of the housing market does not solely affect the real estate sector, but many other sectors which provide numerous services and products.

Previous research suggests that housing price trends were affected by many different factors. For example, Lee and Reed (2014)'s research suggests that the First Home Owner Grant scheme stabilized the housing price in Australia. In Beijing, the three market fundamentals – namely, the population, the housing vacancy area and urban disposable income – explained 60% of the observed housing prices changes (Li and Chand 2015). In Hong Kong, capital market variables explain 75% of the changes in housing prices (Chau *et al.* 2001a). Furthermore, residential units with lucky floor numbers are more expensive during property booms but not in property slumps (Chau *et al.* 2001b). Another piece of research found that an increase of one millimeter's annual rainfall led to a decrease in housing prices from £4 to £14 per square meter. Home purchasers living in regions with higher rainfall were willing to pay less for an extra millimeter of rain on average (Li 2009).

Housing markets are also affected by economic and financial conditions. Various macroeconomic factors affect real estate investment yields and home purchasers' returns. When the economy is bad, unemployment is high, workers scramble for jobs and workers' incomes fall; capability and confidence of home purchase drops substantially, leading to a potent negative drag on the number of homebuyers. Hence, in times of global financial crisis – such as, the Asian Financial Crisis in 1997 and the Global Subprime Financial Crisis – not only did income across the board decline, causing local housing prices to drop substantially, but housing prices in other places also recorded a huge drop. In India, while the share price index, non-food bank credit and foreign direct investment positively affected housing prices, the rate of inflation and market capitalization negatively affected housing prices (Mallick and Mahalik 2015). Capital inflows which were too large to absorb and distorted policy choices have always been regarded as a major factor, which led to the previous real estate sector's collapse (Cheong *et al.* 2014). In Korea, for example, it was found that the impact of school proximity and scenic views dropped after the financial crisis (Kim *et al.* 2015).

As housing prices are quite high in many places, many of us do not have sufficient funds and rely on the mortgage provided by financial institutions such as banks. Many homeowners need to pay interest to these companies periodically. Beyond doubt, interest rates always play an important role in our global housing market. In times of high interest rates, the costs of borrowing are high and the number of borrowers decreases accordingly (Li 2015a). Therefore, many countries and cities increase interest rates to control the skyrocketing housing market, or lower rates substantially to prevent the collapse of housing prices.

Similar to many other products in the market, various characteristics of housing affect the changes of property prices. The nature of the property, quality of tenant, liquidity, duration of the lease, location and nominal yields are important factors which also affect property market yields and demand for home-ownership (Karlsson, 2003). In some highly populated cities, such as Hong Kong, correlations between the business cycle and the housing price cycle are very high. Some researchers have even commented that the "housing cycle *is* the business cycle" (Leamer 2007).

To study the implications of various macroeconomic factors on the housing market, much research has applied various econometrics models to the factors that drive the ups and downs of demand, supply and prices in various housing markets around the world; models such as multivariate time series models, State Space models, Vector Error Correction models, and so on. Despite the fact that econometrics techniques have been applied in housing market analysis, academic research which has applied econometric models to international housing markets is scarce. One of the objectives of this book, therefore, is to provide information on how to apply econometric techniques to different housing markets around the globe.

Is housing market analysis a local or an international issue? Due to the immobile nature of housing, housing research has traditionally been local. The provision of local facilities – proximity to the public transportation network, shopping malls, schools, libraries, swimming pools – is often identified as an important factor which drives or reduces the demand for housing and housing prices. Likewise, local environmental factors such as water seepage (Li and Li 2011, Li 2012), air quality (Chau *et al.*, 2005 and Chau *et al.*, 2011), and noise level (Chau *et al.* 2005) also affect housing prices. For practitioners, local knowledge is very important in making informed housing investment decisions.

From the early 1990s, technological breakthroughs have changed the façade of technology. The development of information technologies not only changes our form of communication but also shortens the distance between different parts of the world. We know more about cities that are far away from us through the World Wide Web. Today, we can acquire information on the housing market in different cities from various real estate resources, such as websites provided by the government, real estate consultants, agents and developers through the Internet. Knowledge of the housing market is no longer local.

In the twenty-first century, local housing market information is globally available to Internet users. We can be a knowledge receiver and sharer at the same time. In addition, the popularity of social media such as Twitter, Facebook and LinkedIn has changed to norm of communications, information distribution and knowledge sharing process. The use of the smart phone even opens up another gateway of communication and knowledge sharing. Many of us now use Web 2.0 tools, such as WhatsApp, YouTube and WeChat, to communicate and share what we know (Li 2015b, Li and Poon 2013, Li and Ah Pak 2012). Despite the potential danger of information explosion and overload, it is undeniable that the Internet has tremendously reduced the transaction cost of obtaining housing market information (Springer 1996). At the same time, the availability of housing market data has made quantitative analysis of the housing market using econometric tools more viable than before. Housing researchers and analysts need to use these tools to make the best of available housing market information.

As a result of globalization, there is an increasing proportion of non-local players in the housing market, especially in markets with little restriction on capital flows. In many cities – such as, New York, London and Hong Kong – the percentage of international buyers in the housing market has been increasing. Globalization has also led to more independence of economic performance across different regions. Since the housing market is closely related to economic performance, regional housing markets today cannot be viewed as closed and independent of other markets.

In view of the above, real estate researchers, consultants and investors need to have a regional and global perspective in order to better understand the housing market, as well as the skills to analyze market data, which is now much more readily and cheaply available. Taking examples from various housing markets to illustrate the application of econometric tools, this book introduces readers to the variety of housing markets around the world. In this book we include studies of mainland China, Hong Kong, South Africa, the Czech Republic, Germany, Norway, Canada, Japan, the US and the UK.

Chapter 2 provides a brief introduction of the role of econometrics modeling in global real estate market research. It provides a summary of the mathematical background of State Space models, the HP filter, VECM, impulse response functions, Probit model, Hedonic Pricing models and Cobb–Douglas models. Studies of these techniques will be used to illustrate their application.

Chapter 3 studies various factors which affect housing entrepreneurs' decisions in China's housing market. Traditional economists consider the role of the entrepreneur as being the one who bears the risk. Successful entrepreneurs are also the group of elites who reap a profit in return for risk-taking and the exercise of initiative. Housing developers are no different. In many countries, developers predict the taste of potential buyers and make decisions on the basic fittings (floor and wall finishes, bathroom fittings, kitchen cupboards) so as to save the buyer the time of decorating the housing unit before they move in. Developers can also provide the fitting and finishes more cheaply due to economies of scale. In mainland China, however, some units or houses are sold without floor or wall finishes, kitchen or bathroom fittings (bare units). What is the motive behind such a choice? This chapter analyses the factors which affect housing developers' decisions to provide fittings by looking at 1,701 first-hand housing developments in Chongqing and Hangzhou by using a dichotomous Probit model. The results show that developers build a higher proportion of bare units in mainland China when: (1) there is a shortage of supply; and (2) land costs are high, so the costs of providing fittings and finishes become relatively low.

Apart from the direct housing market, it is often thought that real estate is an important indirect housing investment, which allows small investors with limited capital to invest in the physical market. Chapter 4 sheds light on the methods that can be used to forecast real estate stock prices. Previous studies show that forecasting stock prices is difficult because stock price performance is affected by many factors, both the macro economy and the micro operation of companies. Furthermore, data is usually reported at varying frequencies. While some are reported monthly, others are released semi-annually. This chapter aims at finding the relationship among the macro economy, corporate performance and companies' stock prices. In this chapter, we forecast the monthly property stock price using the State Space model, which has its origin in engineering. It helps us to predict the monthly stock price by using lower frequency macroeconomic data, often released yearly or half-yearly, as well as biannual micro corporate performance data.

Many finance theories were built on the assumption of rational human behavior and negligible transaction cost. They assume that asset markets are efficient: market participants maximize returns on investment by making use of all the available information that will affect an asset's price. This means that any observed asset price reflects all available information, which renders forecasting of future asset prices impossible. Nevertheless, the high transaction costs in the housing market implies that the capitalization of price-influencing information into asset prices may be a slow process, and therefore future prices may be predictable. The 2013 Economic Nobel Prize laureate Eugene Fama suggests that asset prices are unpredictable. However, another 2013 Nobel Prizewinner in Economics, Robert Shiller, proposes that the rises and falls in asset prices were often guided by the psychology of the investors: ups and downs of the asset prices could be guessed at by studying market investors' behavior. Given the above two diverse but contradictory points of view, we look at property market behaviors in times of peak and trough via news recorded in 2003 and 2013 in Hong Kong in Chapter 5.

Many Hong Kong Chinese have an article of faith in lucky numbers (Chau *et al.*, 2001b). They believe that the pronunciation of four and 14 are similar to the Chinese word "dead." The number eight resembles the idea of wealth while three means "alive," which brings good luck and health to residents. Hence, many of the modern designers of residential estates remove all the so-called unlucky floor and room numbers. Some firemen, however, find this adversely affects their work: when somebody claims to be on the 63th floor, that may instead be on around the fortieth floor only. Chapter 6 sheds light on Tsueng Kwan O, a new town in Hong Kong with many young couples and families, to test if there is any impact of lucky and unlucky numbers on residential property prices, and indirectly on the younger generation.

Previous research suggests that positive externalities lead to higher property prices while negative externalities lead to lower property prices. Nevertheless, few studies have been conducted on the impact of better views, air and noise pollution on housing prices. Chapter 7 studies the impact of positive and negative externalities from 1994 to 2013 on property values in Amoy Garden, one of the largest properties in Hong Kong.

We shed light on the impact of better views, noise pollution and air pollution on the property values. All the Hedonic models show that residential units on lower floors sold at lower prices, though flats which faced the noisy road sold at higher prices. Furthermore, property values of flats on lower floors dropped greater than those of higher floors if the air pollution became worse. This implied that the positive impact of a view outweighed the negative impact of air and noise pollution.

As previous research showed that financial crises usually impose negative impacts on exports and unemployment rates, Chapter 8 reviews the subprime mortgage crisis problem in the US from 2007 to 2009. It was considered the worst financial crisis since the Great Depression. This chapter analyzes the impact of the subprime financial crisis before and after the financial crisis with the help of Chow tests, L1 model and Quantile regression analysis. The results showed that housing prices in Germany dropped during the subprime financial crisis as shown in Quantile regression and Chow tests, although the impact on Norway's housing prices was insignificant during the whole period of the financial crisis.

Chapter 9 studies the interest rate exposure of the housing markets and the role of housing prices in the monetary transmission mechanism in two emerging markets, the Czech Republic and South Africa. The Granger causality test results indicated that housing price fluctuations create wealth and a balance sheet effect in both countries. The results of impulse response functions based on the Vector Error Correction models showed that the impact of wealth and the balance sheet effect are greater in South Africa, and South Africa faced a greater interest rate exposure in housing.

Similar to many developed countries, such as the United States, the United Kingdom, Germany and Spain, Canada has a well-developed financial and housing market. In Chapter 10, we study the various factors which might affect the housing prices in Canada according to various macroeconomic and housing finance inputs. We first provide a comprehensive review of Canada's macroeconomic background, demographic information as well as the residential mortgage market in Canada. After that, the Cobb–Douglas Model based on bootstrapping data is used to examine the factors which affect housing prices.

Apart from direct markets, indirect real estate – such as the Real Estate Investment Trust (REIT) – allows investors with limited funds to invest in the real estate market. It is a valuable tool for global investors who may want to invest in overseas real estate markets but do not wish to invest in physical forms, due to shortcomings such as indivisibility, illiquidity, lack of local knowledge, high information and transaction costs. REITs' underlying assets include housing, hotels, retail and so on. In Chapter 10, we aim to reveal the correlations between REITs cycles in four major stock markets: Hong Kong, Japan, the US and the UK. We use MCMC to simulate the missing data. The HP filter decomposes the time series data into cycle and trend. Pearson correlation analysis shows that the correlation between REITs cycles are positive and significant. Investors need to consider the ups and downs of local and foreign markets. This is the first research which studies the correlation of REITs cycles in four places by using MCMC, HP filter and Pearson correlation. Finally, we conclude the book by using the 'Wh' questions in Chapter 11.

All in all, this book offers insights and reflects the interplay of various economic factors and housing prices via several econometric models. The authors hope that academic faculty members, real estate/economics students, real estate analysis and consultants find the information in this book useful.

Acknowledgement

The authors would like to thank the reviewers of the book who offered constructive comments and suggestions to improve its contents. In particular, we are indebted to Helena Hurd and Sade Lee of Routledge for their support and patience throughout the lengthy process of producing this book.

References

Chau, K. W., B. D. MacGregor and G. M. Schwann (2001a). Price Discovery in the Hong Kong Real Estate Market. *Journal of Property Research*, 18(3), 187–216.

Chau, K. W., V. S. M. Ma and D. C. W. Ho. (2001b) The Pricing of "Luckiness" in the Apartment Market. *Journal of Real Estate Literature*, 9(1), 29–40.

Chau, K. W., F. F. Ng and E. C. T. Hung (2001c) Developer's Goodwill as Significant Influence on Apartment Unit Prices. *Appraisal Journal*, 69, 26–30.

Chau, K. W., S. K. Wong and C. Y. Yiu (2005) Improving the Environment with an Initial Government Subsidy. *Habitat International*, 29(3), 559–569.

Chau, K. W., S. K. Wong, A. T. Chan and K. Lam (2011) "The Value of Clean Air in High Density Urban Areas." In *High-Rise Building Living in Asian Cities*, first edition, (A. G. O. Yeh and B. Yuen ed.), Springer Verlag, 113–128.

Cheong, K. C., P. P. Lee and K. H. Lee (2014) Developers and Speculators: Housing, Ethnic Chinese Business and the Asian Financial Crisis in Malaysia. *Journal of Contemporary Asia*, 44, 616–644.

Karlsson, B. (2003) Property Yields – Concepts, Determinants and Measurement Problems. *Journal of Buildings and Real Estate Economics*, 218, 42–44.

Kim, H., S. W. Park, S. Lee and X. Xue (2015) Determinants of House Prices in Seoul: A Quantile Regression Approach. *Pacific Rim Property Research Journal*, 21, 91–113.

Leamer, E. E. (2007) Housing IS the Business Cycle. *NBER Working Paper No. 13428*, 2–74.

Lee, C. L. and R. Reed (2014) The Relationship between Housing Market Intervention for First-Time Buyers and House Price Volatility. *Housing Studies*, 29, 1073–1095.

Li, Q. and S. Chand (2015) Market Fundamentals, Rational Expectation and Housing Price Changes: Evidence from the Housing Market in Beijing. *Housing, Theory and Society*, 32, 289–301.

Li, R. Y. M. (2009) The Impact of Climate Change on Residential Transactions in Hong Kong. *The Built and Human Environment Review*, 2, 11–22.

Li, R. Y. M. and Y. L. Li (2011) Judges' View on the Price of Environmental Externalities in the United Kingdom. *US–China Law Review*, 8, 994–1007.

Li, R. Y. M. (2012) The Internalization of Environmental Externalities Affecting Dwellings: A Review of Court Cases in Hong Kong. *Economic Affairs*, 32, 81–87.

Li, R. Y. M. and D. H. Ah Pak (2012) Strategic Universities Course Management in Knowledge Explosion Age. *International Journal of Information Processing and Management*, 3, 26–36.

Li, R. Y. M. and S. W. Poon (2013) *Construction Safety*, Springer, Berlin.

Li, R. Y. M. (2015a) "Construction Safety Knowledge Sharing via Smart Phone, Handbook of Mobile Teaching and Learning." In *Handbook of Mobile Teaching and Learning*, Ed. Aimee Zhang, Springer, Berlin.

Li, R. Y. M. (2015b) Generation X and Y's Demand for Home-Ownership in Hong Kong. *Pacific Rim Property Research Journal*, 21, 15–36.

Mallick, H. and M. K. Mahalik (2015) Factors Determining Regional Housing Prices: Evidence from Major Cities in India. *Journal of Property Research*, 32, 123–146.

Sirmans, S., D. Macpherson and E. Zietz (2005) The Composition of Hedonic Pricing Models. *Journal of Real Estate Literature*, 13, 1–44.

Springer, T. M. (1996) Single-Family Housing Transactions: Seller Motivations, Price, and Marketing Time. *Journal of Real Estate Finance and Economics*, 13(3), 237–254.

2 Applied econometric models in international housing markets

Theories and applications

Rita Yi Man Li, Kwong Wing Chau,
Tat Ho Leung and Li Meng

2.1 Introduction

Applied econometrics has been used to study various macroeconomic conditions, such as construction accident compensation data (Borooah *et al.* 1998), economic growth (Li and Ng 2013), business activities, unemployment rates, the relationship between carbon dioxide emission and GDP (Li and Hung 2013), etc. Another useful application falls on the international housing market. Many applied econometrics models are used to study the relationship between various factors that are related to housing prices. In this chapter, several econometric models are discussed: the Hodrick–Prescott (HP) filter, the Vector Error Correction model (VECM) and the impulse response function, the Hedonic Pricing model and the Cobb–Douglas production function, the State Space model (SSM) and the Probit model. We shed light on the theoretical meanings and important features of the aforementioned econometric models.

2.2 Hodrick–Prescott (HP) filter

Time series data consists of trend and cycle components. The HP filter has been applied in international housing markets analyses for a couple of decades. It performs de-trend algorithms for time series analysis which are similar to Baxter and King (1999) and Christiano and Fitzgerald (2003). These filters were used to separate a time series into cycle and trend. The history of the HP filter can be traced from Hodrick and Prescott (1981), who developed the HP filter to separate a time series y_t to a cyclical component c_t, and a trend component, which is also known as a growth component τ_t, such that:

$$y_t = \tau_t + c_t$$

where t = 1, 2, . . . ,T.

If τ_t is non-stationary it has a deterministic or stochastic trend. HP filter is then used to extract a stationary cyclical component c_t, which is driven by stochastic cycles within a specific period of time (Hodrick and Prescott 1981). In short, it can be represented by the following equation:

$$\tau_t = y_t - c_t$$

The HP filter is a trend-removal technique which is applied in macroeconomic analysis, such as business cycle and gross domestic product trend. This mathematical tool decomposes the cycle and trend components from a time series data series to achieve different research aims and objectives. The HP filter applies parameter (λ) to derive a smooth trend. The smoothening has a positive relationship with λ such that $\lambda \to \infty$ leads to a smooth trend. However, it can be controversial with regards to the exact value of λ in different frequencies (Hodrick and Prescott 1981). In general, Hodrick and Prescott (1981) recommended 1600 as the value of λ in quarterly data.

The measure of smooth $\{\tau_t\}$ is the sum of the square of its second difference. c_t denotes the deviations from τ_t in long time periods such that their average is close to zero. The HP filter is obtained by solving the following mathematical function (Hodrick and Prescott 1981):

$$\min_{\tau_t}\left(\sum_{t=1}^{T} c_t^2 + \lambda \sum_{t=1}^{T} \left(\left(\tau_t - \tau_{t-1} \right) - \left(\tau_{t-1} - \tau_{t-2} \right) \right)^2 \right)$$

where $c_t = y_t - \tau_t$.

If λ is sufficiently large, all the $\tau_{t+1} - \tau_t$ will be a constant at an optimum level, β, such that $\tau_t \cong \tau_0 + \beta_t$. It implies that λ converges to the least square of a linear time trend when it approaches to infinity (Hodrick and Prescott 1981). According to Hodrick and Prescott (1981), λ smoothens the time series by penalizing the trend component τ_t. Hence, the first order condition for τ_t shall then become:

$$\tau_t = \frac{y_t}{\left[\lambda L^{-2} \left(1 - L \right)^4 + 1 \right]}$$

where L denotes the lag operator.

Since $c_t = y_t - \tau_t$, the cyclical component, $c_t = HP(L)\, y_t$ will become:

$$HP(L) = \frac{\lambda L^{-2} \left(1 - L \right)^4}{\lambda L^{-2} \left(1 - L \right)^4 + 1}$$

According to Hodrick and Prescott (1981), when $\lambda \to \infty$, τ_t approaches a linear time trend. Alternatively, the above formula can be rewritten in the following matrix form:

$$(Y - \Gamma)' \, (Y - \Gamma) + \lambda \, \Gamma' \nabla^2' \nabla^2 \Gamma$$

where Y and Γ are T $\times$ 1 and represent the vectors of the time series and trend. ∇^2 denotes the second difference matrix. The solution can be obtained from the first order condition:

$$\Gamma = (I + \lambda \nabla^2' \nabla^2)^{-1} \, Y$$

$$C = Y - \Gamma$$

where C is T $\times$ 1 vector from the cyclical component.

Moreover, there are different ways to modify the HP filter to satisfy the needs of researchers for different research purposes. For example, Ravn and Uhlig (2002) determined λ under different frequencies of observations.

2.2.1 HP's application in housing market analysis

Witkiewicz (2002) applied the HP filter to analyze the Swedish real estate cycle. Witkiewicz first performed a stationary test to ensure that the HP filter was symmetric, which would not create artificial cycles as suggested by (Ahumada and Garegnani 1999, Park 1996). The author then filtered the cycle out by normalizing the data series via the traditional method:

$$S_i^N = \frac{S_i - \overline{S}_i}{\sigma_i}$$

where $\overline{S}_i$ and σ_i denoted the average of data series and standard deviation respectively. The analysis showed the lead–lag relationships among the estimated variables such that the cycle should be a function of the separate HP filtered variable series, such that:

$$\text{Indicator} = f(HP(L))_t = w'L$$

where w' and L denote the row vector of weights and column vectors of a HP filtered variable series. It should be noted that there are other methods to construct the cycle component by using the actual time series data without normalization and may lead to different results.

2.2.2 Shortcomings of the HP filter

Similar to many other research methods, however, the HP filter is not perfect: it has its restrictions and shortcomings. For example, French (2001) indicated that the HP filter only reached optimal results under two restrictive assumptions. First, the data is in I(2) trend, or shifts in trend of growth rates do not exist in raw data. King and Rebelo (1993) and Cogley and Nason (1995) similarly thought that another approach should be used, instead of the HP filter, if there are one-time permanent shocks or significant split trend growth. Second, French (2001) stated that white noise is a necessary condition to apply the HP filter. It is optimal only if the analysis is in a closed domain with historical and static data as it misled the predictions in dynamic data.

Baxter and King (1999) proposed similar results for λ = 1600 and suggested that the Band-Pass filter should have a cutoff value at 16 or 32 quarters. They concluded that the poor performance of the HP filter may be the result of a low frequency trend movement.

Furthermore, Stock and Watson (1999) pinpointed that the two-sided HP filter includes the observations at t + i, i > 0 which was used to estimate the current time t. Instead of using a two-sided HP filter, it was suggested to make good use of a one-sided HP filter. The major drawback of the HP filter is the so-called "end point bias." In general, even though the HP filter shares some shortcomings, it is commonly used as a popular analytic tool among researchers. In the following discussion, two examples illustrate the application of the HP filter to the housing market.

2.2.3 Example 1: is there seasonality in housing prices? Empirical evidence from CBSAs

Many product sales are affected by seasonal factors. For example, more ice creams are sold during summer and substantially fewer in winter. There are more heaters sold in winter than summer. Therefore, removing the seasonality problem has become a popular and important application of the HP filter. Miller *et al.* (2013) collected the housing transaction data from 138 Core Base Statistical Areas in the US (CBSAs) from February 2000 to April 2011. It was observed that monthly frequency could not be directly applied by using $\lambda = 1600$ as the parameter.

Miller *et al.* (2013) separated the change in housing price into trend and cyclical components by way of the nested variable regression. The selection of the parameter λ was based on the research results of previous literatures. They adopted the nested annual data from Mcgough and Tsolacos (1995) who suggested that $\lambda = 100$ should be used for annual data. It was tested by using empirical data from the UK. After applying the HP filter, there were 135 cyclical coefficients in the sample period. The average monthly variation over a year showed that the summer months recorded higher sale prices than the other seasons with a peak in June, while the winter months recorded lower prices with a trough in January. It was suggested that the range of variation was almost 3%, from the highest to the lowest.

After applying the HP filter, the trend component was extracted. Inspired by (Goodman 1993, Kaplanski and Levy 2012), Miller *et al.* (2013) included geographic regions, tourism, ethnicity and weather variations as important factors to study the home prices changes due to seasonality problem. The results showed that only tourism was statistically significant in these two regions. There was a lower variability in price in western regions than eastern and higher variation in the Mid-west region than the east.

2.2.4 Example 2: asset price boom and bust cycles analysis using the HP filter

The HP filter can also be used to detect the existence of bubbles. Since it extracts a trend, extremely high or low prices can be easily detected. In general, it extracts the trend from the time series housing price index data. The booms and busts can be easily found in graphical form. The filter defines a bubble as an abnormal time period with larger than δ times its standard deviation (Borgy, Clerc and Renne 2009).

The HP filter was applied in many of the previous research (Goodhart and Hofmann 2008, Cecchetti 2006). In addition to the traditional HP filter applications, there are two other modified versions of the traditional HP filter brand, which is known as the Extended HP (EHP) filter, and the Recursive HP (RHP) filter. The EHP filter aims to alleviate the problems in the typical HP filter. First, it smoothens the series into quarterly frequency. Second, it extends the time period by adding n quarters before the date of a pre-determined threshold. The RHP filter is another way which detects episodes that are longer than the typical HP filter. It was applied in previous research (Adalid and Detken 2007, Alessi and Detken 2009, Detken and Smets 2004). Indeed, the RHP filter is a popular method that many researchers use to reduce end-point bias.

Borgy *et al.* (2009) estimated the real housing quarterly index from 1970 quarter one to 2008 quarter three in 18 OECD (Organisation for Economic Co-operation and Development) countries.[1] They compared the EHP filter and RHP filter and found that these results shared similar conclusions.

After checking the coherence of the filter and index series, the probability of "a boom is followed by a bust" was estimated. It was found that there was a 50% chance of a bust in the next period in the EHP filter methodology.

As a matter of fact, the HP filter can be applied to many different scenarios. In particular, it works well with cyclical analysis (such as business, construction, and stock cycles) and trend analysis (such as gross domestic product trend). The major concern of the HP filter application will be which type of HP filter should be chosen as well as the value of the parameter.

2.3 Vector Error Correction model (VECM) and impulse response

Sometimes, changes in economic or financial situations scoop up some houses at a huge discount. We then analyze the shocks of such a change in housing prices.

Table 2.1 The results of moving average, EHP filter and RHP filter (Borgy *et al.* 2009)

	Extended HP filter	*Moving average*	*Recursive HP filter*
Moving average	–	–	–
HP filter (recursive)	–	59%	–
HP filter (extended)	–	53%	64%
Band-pass filter	75%	53%	64%

1 OECD member countries are Australia, Belgium, Canada, Denmark, Finland, France, Germany, Ireland, Italy, Japan, the Netherlands, New Zealand, Norway, Spain, Sweden, Switzerland, the United States, the United Kingdom.

Table 2.2 Probabilities that an asset price boom was followed by a bust (Borgy *et al.*
2009)

Probabilities of an asset price boom followed by a bust	Housing prices	Stocks prices
Recursive HP filter		
Whatever type of boom	23%	9%
If the boom is not costly	18%	0%
If the boom is costly	28%	31%
Extended HP filter		
Whatever type of boom	47%	57%
If the boom is not costly	50%	52%
If the boom is costly	45%	64%
Band-pass filter		
Whatever type of boom	35%	67%
If the boom is not costly	22%	52%
If the boom is costly	41%	84%
Moving average		
Whatever type of boom	67%	70%
If the boom is not costly	60%	72%
If the boom is costly	74%	65%

Impulse response can be used for analysis as such. In general, impulse response refers to the reaction to an external shock in a dynamic system. In econometrics, impulse response is a popular method for studying the impact of market fluctuation and sudden change (often refers to a shock). It is usually adopted in financial and stock market analysis and is often the final step in VAR or VECM analysis. It can be applied to examine the dynamic relationships between a bundle of stocks in the same or different industries. If there are price changes due to shocks, impulse response shows how the other stocks react. It also examines how long it takes to react to the prices changes due to shocks and sudden changes. Apart from stock market analysis, impulse response can also be applied in housing prices analysis. There are both direct and indirect impacts which are reflected in the housing market from such financial fluctuations. Therefore, it can be used to estimate the various shocks on housing markets.

Li and Ng (2013) applied a VAR model to study the relationship between FDI and real GDP growth in South Africa. Coulson and Kim (2000) applied a multivariate VAR with impulse response to test the proposition that residential investment has considerable impact on macro economy (GDP). Koop *et al.* (1996) developed the impulse response in a nonlinear multivariate model. Later, Pesaran and Shin (1998) revised and extended Koop *et al.* (1996)'s traditional impulse response model to the generalized impulse response model. The impulse response function gives the *jth* period response while the system receives a shock with one standard deviation. Suppose a shock $\{\bar{\varepsilon}_t\}_{t=1}^{\infty}$ occurred:[2]

$$y_t = \rho y_{t-1} + \varepsilon_t, \varepsilon_t \sim N(0,\sigma^2)$$

The generated time series y_t will then become $\{\bar{y}_t\}_{t=1}^{\infty}$. Meanwhile, the shock will also affect ε_t, such that the generated series will become:

$$\bar{\varepsilon}_t = \begin{cases} \bar{\varepsilon}_t + \sigma & \text{if } t = \tau \\ \bar{\varepsilon}_t & \text{o.w.} \end{cases}$$

Furthermore, the concept of its function can be formalized as:[3]

$$Y(j) = E\left[y_{t+h} \mid u_{t+j} = \begin{cases} d_i & \text{if } j = 0 \\ 0 & \text{if } j \in (1,h) \end{cases}; I_t \right] - E\left[y_{t+h} \mid u_{t+j} = 0, \forall j \in (0,h); I_t \right],$$

such that this function is used to estimate the response of the system at time, t + h, for h = 0, . . . , H due to an existence of shock from the disturbance n × 1 vector, $d_i \cdot I_t$ refers to the information available at time t. The disturbance vector depends on the number of dimensions. To simplify the issue, impulse response function can alternatively be presented as:

$$Y(j) = \tilde{y}_{\tau-1+j} - \bar{y}_{\tau-1+j}$$

The linear process of the impulse response function application starts at the steady state where $\bar{\varepsilon}_t = 0$ such that

$$Y(j) = \sigma\rho^{j-1}$$

Although a general formula can be generated for the impulse response function process, it depends on the model in different scenario. The following equations illustrate a hypothetical example that can be used to understand the real life application.

Consider a simple bivariate VAR(1) model:

$$y_{1t} = \beta_{10} + \beta_{11} y_{1t-1} + \alpha_{11} y_{2t-1} + u_{1t}$$
$$y_{2t} = \beta_{20} + \beta_{21} y_{2t-1} + \alpha_{21} y_{1t-1} + u_{2t}$$

---

2 This framework was first developed by W. Enders (2004) in *Applied Econometric Time Series*. John Wiley & Sons, Inc.

3 This following part of framework is largely inspired by D. Ronayne (2011), Which Impulse Response Function? In *Warwick Economic Research Papers*, Department of Economics, The University of Warwick.

These equations demonstrate how the changes of one single unit or part of the economic system will lead to different responses in other series. Suppose there is a unit shock in u_{1t} with the immediate effect y_{1t} in response to the changes. At time $t + 1$, y_{1t} will change to y_{1t+1} and y_{2t+1}. In general, the shock would affect the decreasing magnitude until it reaches the new steady state, i.e. the new equilibrium where it is suffice to say that the shock has been completely absorbed. In fact, there are some conditions that the shock could not be absorbed. It entirely depends on the estimators in the models or the estimator matrix.

Suppose the above bivariate VAR(1) model is presented in a matrix form with a unit shock to y_{1t} at time $t = 0$.

$$y_t = A_1 y_{t-1} + u_t$$

where A_1 is a 2×2 matrix. In general, A_j is an $n \times n$ matrix.
Thus, it is now presented as the following equation:

$$\begin{pmatrix} y_{1t} \\ y_{2t} \end{pmatrix} = \begin{pmatrix} \beta_{11} & \alpha_{11} \\ \beta_{21} & \alpha_{21} \end{pmatrix} \begin{pmatrix} y_{1t-1} \\ y_{2t-1} \end{pmatrix} + \begin{pmatrix} u_{1t} \\ u_{2t} \end{pmatrix}$$

First, at time $t = 0$,

$$y_0 = \begin{pmatrix} 1 \\ 0 \end{pmatrix}$$

Second, at time $t = 1$:

$$y_1 = \begin{pmatrix} y_{11} \\ y_{21} \end{pmatrix} = \begin{pmatrix} \beta_{11} & \alpha_{11} \\ \beta_{21} & \alpha_{21} \end{pmatrix} \begin{pmatrix} 1 \\ 0 \end{pmatrix} = \begin{pmatrix} \beta_{11} \\ \beta_{21} \end{pmatrix}$$

Third, at time $t = 2$:

$$y_2 = \begin{pmatrix} y_{12} \\ y_{22} \end{pmatrix} = \begin{pmatrix} \beta_{11} & \alpha_{11} \\ \beta_{21} & \alpha_{21} \end{pmatrix} \begin{pmatrix} \beta_{11} \\ \beta_{21} \end{pmatrix} = \begin{pmatrix} \beta_{11}\beta_{11} + \alpha_{11}\beta_{21} \\ \beta_{21}\beta_{11} + \alpha_{21}\beta_{21} \end{pmatrix}$$

Although it is possible to compute y_n, it is not the general formula used for impulse response analysis. After obtaining the values of $(y_0, y_1, \ldots)$, the impulse response diagram can be plotted. The following shows two examples to illustrate the application of impulse response function in housing research.

2.3.1 Example 1: residential investment and business cycles in an open economy: a generalized impulse response approach applied in the US

Bisping and Patron (2008) studied the impact of residential and non-residential investment on economic growth in the US. Inspired by Pesaran and Shin (1998), Bisping and Patron (2008) used a generalized impulse response function to study the impact of residential and non-residential investment. They collected the quarterly data of consumption, government expenditures, residential investment and non-residential investment quarterly data from 1959 quarter one to 2004 quarter four. After that, they applied the VECM method to conduct impulse response analysis. According to Bisping and Patron (2008), the following empirical model was used to estimate the impact of the various factors:

$$x_t = \mu + \sum_{j=1}^{T} B_j X_{t-j} + \varepsilon_t$$

Where x_t is a 5×1 vector consisting of $C_t, G_t, RI_t, nRI_t, NX_t$ and represents the data mentioned above. At the same time, μ denotes a 5×1 vector, B_j denotes 5×5 matrices of parameters and ε_t denotes 5×1 vector of error such that variance, $E(\varepsilon_t \varepsilon_t') = \Sigma$ which is a 5×5 matrix in terms of σ_{ij}. Thus, transferring the VAR model into MA(∞):

$$x_t = \mu + \sum_{j=0}^{\infty} W_j \varepsilon_{t-j}$$

Where W_j is a 5×5 matrix of parameters with fulfillment in such conditions: $W_0 = I_5$, where I_5 is 5×5 identity matrix:

$$W_j = B_1 W_{j-1} + \ldots + B_T W_{j-T}, j > 0$$
$$W_j = 0, j < 0$$

Therefore, the generalized impulse response function for x_t with shock in δ_j to the *jth* element of the VAR periods will then become:

$$E(x_{t+s}|\varepsilon_{jt} = \delta_j, I_{t-1}) - E(x_{t+s}|I_{t-1})$$

where E denotes the expectation operator and I_t denotes all the information at time $t - 1$. Since ε_t is multivariate normally distributed, the terminal value will be:

$$\frac{W_s \sum e_j}{\sqrt{\sigma_{jj}}} \frac{\delta_j}{\sqrt{\sigma_{jj}}}$$

where e_j is a 5×1 vector that $= 1$ in the *jth* element, otherwise, $= 0$.

It was concluded that although both the residential and non-residential investment affected GDP, residential investment has a stronger impact in the early periods. However, net exports dampened the disparity in impact of these two variables. Meanwhile, consumption and government expenditure is consistent throughout the whole period.

The above section has discussed a typical application procedure in VECM analysis. However, there are some further analyses which can be conducted using VECM. In this section, we will discuss one of the most famous forecasting analyses in VECM – impulse responses. Impulse responses are often used to trace out the responsiveness of some of the dependent variables to shocks from an error term. Consider a simple bivariate ECM:

$$\Delta Y_t = \varphi + \delta t + \lambda_{e_{t-1}} + \gamma_1 \Delta Y_{t-1} + \ldots + \gamma_i \Delta Y_{t-i} + \omega_1 \Delta X_{t-1} + \ldots + \omega_i \Delta X_{t-i} + \varepsilon_t$$

$$\Delta X_t = \varphi + \delta t + \lambda_{e_{t-1}} + \gamma_1 \Delta X_{t-1} + \ldots + \gamma_i \Delta X_{t-i} + \omega_1 \Delta Y_{t-1} + \ldots + \omega_i \Delta Y_{t-i} + \varepsilon_t$$

If ε_t has a unit shock occurred at t, it will affect ΔY_t. Meanwhile, it also affects the ΔX_t. Impulse responses are usually represented in graphic form to show the duration required to absorb a unit shock.

After explaining the essential mathematical principle of the VECM, we hereby showcase how VECM works in housing market analysis.

2.3.2 Example 2: forecasting and assessing Eurozone countries housing prices under the lens of the major economic fundamentals

Gattini and Hiebert (2010) forecast the housing prices according to the reduced form model and provided structural decomposition to identify the "common trend" with regards to transition and permanent shock. The VECM is used to analyze the responsiveness of a structural shock. Thus, a VAR model with an assumption of I(1) is established accordingly:

$$\Delta X_t = \mu + D(L)\epsilon_t$$

where $\epsilon_t \sim i.i.d.$ linear forecast errors in X_t.

We can then rewrite the VAR into a reduced form to include a structure, Γ_0:

$$\Delta X_t = \omega + D(L)\eta_t$$

where $\eta_t \sim i.i.d.$ structural disturbances, $\omega = \Gamma_0^{-1}\mu$ and $D(L) = \Gamma_0^{-1}C(L)$ according to Gattini and Hiebert (2010).

The Johansen test should be estimated before constructing the model. Comparing the three criteria, the Johansen test indicates there is one co-integrating equation at a 5% level. Therefore, the VECM can be examined. Furthermore, it also provides the Granger causality test. The results show that F-test statistics cannot reject the critical value from Engle and Granger (1987). It implies that it accepts the null hypothesis which the additional variables do not granger cause X_t^*. In general, the structural shocks can be classified into transitory and permanent effects (King *et al.* 1991). Thus, impulse response can be calculated and tabulated as well.

Accordingly, Gattini and Hiebert (2010) examined three permanent shocks which are:

1   Housing market technology shock;
2   Economy wide technological shock; and
3   Financing cost shock.

Their results showed that it required 28 periods to absorb a unit shock from housing demand (Gattini and Hiebert, 2010).

In general, the results were fair as a prediction with such a sample. The forecast statistics considered a typical mean error (ME), root mean squared error (RMSE) and mean squared error (MSE). It predicted principally that house prices in the euro area would decrease until the end of 2010. The accumulated depreciation in real terms would be 8% in three years. Compared with 1986 to 1991 and 1997 to 2007, it predicted that the recovery cycle would be shortened.

2.3.3 Example 3: the relationship between money and housing – evidence for the euro area and the US

Greiber and Setzer (2007) agree that the recent fluctuations in housing prices were correlated with the loose monetary conditions. They examined both the US and the euro area to confirm the bi-directional correlation between money and the housing market. They collected the time series data in the Eurozone countries from 1986 quarter one to 2006 quarter four. The long-run relationship between standard money demand specifications with housing variables was then estimated by using the following equation:

$$(m - p) = \alpha_0 + \alpha_1 gdp_t + \alpha_2 ir_t + \alpha_3 prop_t + \varepsilon_t$$

where *m, p, gdp and ir* denotes nominal M3, GDP deflator, real GDP and interest rate. Furthermore, *prop_t* refers to the real residential property price index or housing wealth indicator. All the abovementioned variables were in logarithms except the interest rate.

According to the results of the Johansen test, there was one co-integrating equation at the 5% level. Greiber and Setzer (2007)'s VECM results showed that the money demand system in short run is a statistical significant estimator of money growth. Meanwhile, the housing price was statistically significant and positively correlated with households' wealth but not the price. Greiber and Setzer (2007) explained that excess liquidity stimulates an increase in housing demand. When people have more money in their pockets, they are more willing and able to buy. Under the lens of econometrics, the authors thought that the housing surge was stronger in a housing wealth model. In the impulse response function, the response of M3 to a unit shock in house price was positive and significant (Greiber and Setzer 2007).

2.4 The application of the Hedonic Pricing model on housing market analyses

To study the impact of various demand and supply, macroeconomic and financial factors on housing prices, much of the previous research adopted the Hedonic Pricing model in various functional forms, such as semi log. The Hedonic Pricing model can also be expressed as a reduced-form equation which incorporates the demand and supply effect on price. Critical reviews of previous applications of Hedonic Price models in the real estate and housing market can be found in Chin and Chau (2003) and Chau *et al.* (2003).

Malpezzi (2002) suggested that the Hedonic Pricing model arose because there are heterogeneous housing stocks and consumers. Since it can be applied to housing markets and there are many possible implicit variables that are excluded from traditional economic analysis – the environment, internal design, the neighborhood, and so on – the Hedonic Pricing model has become one of the most popular tools in housing prices analysis. In fact, it is also a useful tool to investigate the relationship between housing price and environmental externalities such as different types of pollution.

As early as the 1970s, Rosen (1974) provided a basic framework for Hedonic Pricing model analysis which was then followed and adopted by the later research:

$$P_i = f(S_i, N_i)$$

where P_i, S_i and N_i denote housing price, vector of property's structural characteristics and vector of neighborhood characteristics.

Theoretically speaking, the Hedonic Pricing model includes bundles of attributes which affect the property prices. They include a number of demand and supply factors, market structure, preferences and technology (Ekeland *et al.* 2002). Practically, the Hedonic model is used to estimate the marginal contribution of individual characteristics (Sirmans, Macpherson and Zietz 2005). On top of that, the functional form of the Hedonic model should be flexible and depends on the research objectives and aims. Many researchers, such as Cebula (2009), suggest a logarithmic transformation of the dependent variable, which can help eliminate some of the problems of heteroscedasticity.

Apart from the estimation of various factors which drive housing prices, Hedonic Pricing models were used to construct quality adjusted real estate price indices (Chau *et al.* 2005) and also for estimation of the shadow price of housing components. For instance, as the single family homes in the same district share similar structural characteristics, the weight of each attribute reflects purchasers' preferences or taste (McGreal and Paz 2013). Another example is the estimation of people's preference for clean air using hedonic price model (Chau *et al.* 2005 and Chau *et al.* 2011).

2.4.1 Factors which had been included in the Hedonic Pricing model

While it is flexible and easy to construct a Hedonic model with the various econometric or statistics software such as EView, SPSS, Stata, Gretl, we have to think carefully which factors to include in the model from the range of choices. Sirmans *et al.* (2005) suggested eight basic categories. They include construction and structure, internal features, external features, environment (landscape), environment externalities (pollution), facilities, location, transaction related factors and financial issues. In another school of thought, Tekel and Akberishahabi (2013) proposed there were at least three categories of factors which should be added to Hedonic Pricing model analysis: structural, locational and neighborhood (Li *et al.* 2014).

Li (2009) included the climate data in Hong Kong to study the impact of weather on property prices. Monthly property transactions data was collected from the Hong Kong Statistical Department; average temperature and humidity from Hong Kong Observatory in 1997–2006 was included. The results showed that there was a negative and significant relationship between temperature and property prices.

2.4.2 The advantages and limitations in the Hedonic Pricing model

The Hedonic model is a popular method in housing market analysis as compared with other models mentioned in this chapter. There are some reasons for its popularity. For example, this model is flexible. It can include many different possible interactions of events (such as financial/political crisis, policy changes), environmental (such as sea view, mountain view, airport noise, smelly landfill), social factors (such as race, demographic information, facilities, clubhouse, schools, transportation), characteristics of the property (such as floor level, scale of property, size, quality of various fittings) and other relevant factors. More importantly, the model is quite simple. It is quite easy to understand and apply as compared to other econometric models.

2.4.3 Example: Hedonic Pricing model
for housing in Los Angeles and San Diego

Laurice and Bhattacharya (2005) used the Hedonic Pricing model to show the relationship between various factors and housing prices. Based on more than 18,000 observations (per county) in southern California Counties, Los Angeles Counties and San Diego Counties from January 2002 to June 2003, the Hedonic Pricing model was constructed according to the framework provided by Rosen (1974).

It is common to use the semi-log function in dependent variables (housing price) which can also be found in the literature (Goodman and Thibodeau 2003, Rush and Bruggink 2000, Wolverton and Bottemiller 2003). However, in the independent variables, there are still some other functional forms that have been used, such as linear form, cubic form and quadratic form. In this case, the selection of functional form depends on the adjusted R^2. Thus, the empirical Hedonic model was established by Laurice and Bhattacharya (2005):

$$\ln P_i = \alpha + \beta S_i + \gamma LOC_i + \delta(TREND) + \theta(SEASONAL) + \mu(MORT) + \varepsilon_i$$

where $\ln P_i$ denotes the housing price in natural logarithm. i refers to the number of houses sold in the sample period. S_i, LOC_i, $TREND$, $SEASONAL$ and $MORT$ denote the vectors of housing characteristics, location index in cubic form, time trend in cubic form, seasonal dummy variables in cubic form and 30-year fixed conventional mortgage rate respectively.

In Los Angeles County, a spa can led to an increase the price by 4.6%. Similarly, Orange County and San Diego County share a similar positive relationship that the provision of a spa increases the price by 2% and 4.4% respectively. Furthermore, there is only a small impact if cubic and quadratic coefficients are used.

2.5 Cobb–Douglas function for input–output analysis

In addition to the other econometric models, the Cobb–Douglas function is often used to study the input output relationship between various production factors such as capital and labor, the rate of production and elasticity of substitution. The model dates back to the 1920s, when it was used to study the theory of production. It offered an explanation for the major determinants which affected the level of output. The amount produced, was the results of various inputs which led to the changes in the level of output. The amount produced depended upon the quantities of the factors of production employed and the level of technological knowledge. The theory of production provided a vivid account for the level and amounts of factors of production to be employed (Biddle 2012).

In general, the Cobb–Douglas function can be expressed as:

$$Q = f(K, L)$$

However, in production function, there are so many functional forms representing the relationship between three or more variables. Cobb and Douglass (1928) suggested functional form to study the input–output relationship:

$$Y_i = \beta_1 X_{2_i}^{\beta_2} X_{3_i}^{\beta_3}$$

where Y_i denotes the output and X_i denotes input factors.

The input values can be constant returns to scale, increasing or decreasing in scale. For example, Chao (2003) considers the production technology ex ante to be Cobb–Douglas with constant returns to scale, however, as capital goods had already installed, production possibilities was assumed to be in Leontief form as there was no substitutability of energy, labor and capital ex post.

If we take log on both sides, the Cobb–Douglas function can represent in another form,

$$\ln Y_i = \ln \beta_1 + \beta_2 \ln X_{2i} + \beta_3 \ln X_{3i}$$
$$= \beta_0 + \beta_2 \ln X_{2i} + \beta_3 \ln X_{3i}$$

Thus, this linearization process turns the initial non-linear function into a linear function. It should be noted that it was non-linear in X_i but linear in $\ln X_i$. Based on these two forms, there were some economic meanings within the estimator. However, the Cobb–Douglas function is initiated by the pure consideration of mathematical presentation that it is developed without any further concern in economic theory or evidence in production process. Therefore, such economic meanings are supplemented by other economists. Both β_2 and β_3 are the partial elasticity of output (labor and capital). It implies an valuable meaning for further analysis in different fields where β_2 represents the percentage change if there is 1% change in input factor, X_{2i}, holding other input factors constant. Meanwhile, $\beta_2 + \beta_3$ provides the information about the return to scale. If $\beta_2 + \beta_3 = 1$, it implies that it is in constant return to scale which doubles the input factors and would lead to double the output. Therefore, it is more favorable that $\beta_2 + \beta_3 > 1$ which is an increasing return to scale.

After discussing the mechanism of the Cobb–Douglas function, a special example needs to be mentioned. Consider a Cobb–Douglas function as follows:

$$Y_i = \beta_1 X_{2i}^{\beta_2} X_{3i}^{\beta_3} + u_i$$

Since the error term, u_i, is an additional term to the non-linear function. According to this condition, the normal mechanism cannot be changed to linear form. Intriligator *et al.* (1996) managed to fill up this gap by introducing a new Cobb–Douglas function form which is called the Constant Elasticity of Substitution (CES) function:

$$Y_i = A \left[\delta X_{1i}^{-\beta} + (1 - \delta) X_{2i}^{-\beta} \right]^{-\frac{1}{\beta}}$$

where Y_i denotes output and X_i denotes input factors. A refers to scale parameter and δ refers to distribution parameter that $0 < \delta < 1$. β is the substitution parameter. In this CES function, it can convert the non-linear function into linear form no matter which form of stochastic error term exists. However, it is impossible to conduct a CES function with three or more input factors. This is because the distribution parameter will become uncertain if there are more than two input factors.

Practically speaking, unlike the other three methods which have been discussed earlier in this chapter, the Cobb–Douglas function is not used to deal with specific analysis but provides a general function for analysis. It is useful to apply in housing production analysis. Moreover, it is a good opportunity to study construction through the Cobb–Douglas function.

2.5.1 Application to housing market analysis

The Cobb–Douglas function offers a nonlinear form that can further induce ideas of optimization. Therefore, it can be used directly to estimate the costs of tradeoff. In housing market analysis, construction time and cost analysis are the quintessential areas for the application of the Cobb–Douglas function. Some of the previous academic research suggested that the cost function is non-linear, convex or concave (Berman 1964, Lamberson and Hocking 1970, Moussourakis and Haksever 2010).

Such conditions in research are a simple application of economic optimization, or so-called constrained maximization. Moussourakis and Haksever (2010) considered the non-linear time-cost function that analyzes the minimum project cost under budget constraint.

All in all, the Cobb–Douglas function can be expressed as various functional forms. The non-linear form tends to be more realistic as compared to the actual circumstances of increasing labor and capital input, which could imply a diminishing return.

2.5.2 Example 1: Dennis et al.'s approach
to estimate the production function for housing

Dennis, *et al.* (2010)'s research used housing quantities and prices as latent variables to develop a housing production function. This model assumed that the products were homogenous and divisible as denoted by Q. Two factors included land; and a composite of all mobile non-land factors such as machinery and labor were denoted as M and L respectively. Dennis *et al.* (2010) suggested that housing product is non tradable. Thus, the price of house, p_q depended on the location of the housing developments. Based on the mathematical process provided by Dennis *et al.* (2010), the housing production function can be established according to the following function:

$$Q = M^{\alpha} L^{1-\alpha}$$

where $q = m^{\alpha}$. By solving the optimized equations:

$$m\left(p_q\right) = \left(\alpha p_q\right)^{\frac{1}{1-\alpha}}$$

$$s\left(p_q\right) = \left(\alpha p_q\right)^{\frac{\alpha}{1-\alpha}}$$

Thus,

$$v\left(p_q\right) = p_q \cdot s\left(p_q\right)$$

$$= \alpha^{\frac{\alpha}{1-\alpha}} \cdot p_q^{\frac{1}{1-\alpha}}$$

Inverting the function,

$$p_q(v) = \alpha^{-\alpha} v^{1-\alpha}$$

Finally, the linear presentation form will be:

$$r(v) = (1 - \alpha)v$$

In the estimation of the productivity growth in the building industry, Chau and Walker (1988) adopted the Cobb–Douglas production function. Chau and Walker (1994) used the Cobb–Douglas production function to estimate the factors affecting construction labor productivity. Christensen *et al.* (1973) have shown that the Cobb–Douglas Production function is a special case of the more flexible translog production function. The translog production function has been used by Chau and Zhu (2009) to estimate the production function of the construction industry.

2.5.3 Example 2: applications of the Cobb–Douglas production function in construction time ex post cost analysis

The Cobb–Douglas function is popular to study the tradeoff between time and cost. Theoretically, the Cobb–Douglas function carries the characteristic that it can potentially approach to a maximum or minimum value. Therefore, different research has been conducted on the minimization of construction costs compared with reductions in construction time (Feng *et al.* 1997, Ng and Zhang 2008).

Inspired by Shen *et al.* (2012), Hassani (2012) established the Cobb–Douglas function as follows:

$$Q = f(K, L)$$

where *Q, K, L* denote the rate of production output, capital and labor. It can then be formalized as:

$$Q_i = L_i^{\alpha} K_i^{\beta}, \forall i \in S$$

It was suggested that when the construction project is crashed, the rate of production output should be changed according to the level of increment in labor and capital input. At the same time, the cost function is assumed that:

$$TC = c_1 L + c_2 K$$

where TC is total cost. c_1 and c_2 refer to the wage rate and rental rate.

To develop the mathematical model, Hassani (2012) optimized the total cost function with respect to the rate constrain:

$$\min c_{1i} L_i + c_{2i} K_i$$

$$s.t. \quad Q_i = L_i^\alpha K_i^\beta, \forall i \in S$$

After that, an objective function is established:

$$Z_i = w_1 C_i + w_2 T_i$$

where C_i and T_i are normalized scores of costs and construction duration with value falls between zero and one. Furthermore, the minimum value of Z can be represented according to the trend calculated by the weight of cost score. The minimum value of Z implies that the optimal solution can be allocated between construction time and cost.

2.6 State Space models

In housing market analysis, it is common to analyze housing prices by using various econometric models. In fact, the housing bubble phenomenon is interesting for researchers to analyze the price to rent ratio or price to income ratio (Xiao and Tan 2007). To apply the State Space model, we must ensure that the analyzed system is in linear form.

In the State Space model, the internal state of the system establishes a stated equation. The output combines the current system state and current system input for establishing an output equation (Rowell 2002):

$$y = f(x, u)$$

where y and u denote output variables and input variables, x denotes the stated variables and it can rearrange with respect to the system input and current state:

$$x' = g(x, u)$$

Therefore, x' denotes the rate of change in stated variables.

Hamilton (1986) stated that the stated space models capture an observed $n \times 1$ vector, y_t, in terms of an unobserved $r \times 1$ vector, ξ_t, as the state vector for the system. Therefore, the stated equation can be established accordingly:

$$\xi_{t+1} = F\xi_t + v_{t+1}, v_t \sim i.i.d.N(0, Q)$$

where F denotes an $r \times r$ matrix and v_t denotes $r \times 1$ vector. By using recursive substitution, the following equation can be obtained:

$$\xi_{t+m} = F^m\xi_t + F^{m-1}v_{t+1} + \ldots + F^1 v_{t+m-1} + v_{t+m}, \; m = 1, 2, \ldots$$

where F^m denotes matrix, F, multiplied by itself m times. Thus:

$$E(\xi\_(t+m)|\xi\_t, \xi\_(t-1), \ldots) = F^\wedge m\xi\_t$$

It means that the future values of state vector depends on $(\xi_t, \xi_{t-1}, \ldots)$ i.e. the future value can only be estimated according to the current value of ξ_t. If all values of F lie inside the unit circle, the system will be stable.

After that, we construct the observation equation for the system:

$$y_t = A'x_t + H'\xi_t + \omega_t, \omega_t \sim i.i.d.N(0, R)$$

where y_t is the vector assigned above and observed at time t, H' is the coefficient in $n \times r$ matrix and ω_t is the $n \times 1$ vector of measurement error. It should be noted that ω_t is independent from ξ_t and v_τ, $\tau = 1, 2, \ldots$ In addition to the state related terms, there is an $k \times 1$ exogenous vector, x_t, with coefficients, A', in $n \times k$ matrix. Based on this condition, if all the variables are included in the state vector, x_t will be deterministic in theory, however, x_t is usually non-deterministic in practice (Hamilton 1986).

As a matter of fact, the above deduction illustrates only one condition and there are many possible conditions in real life situations. In general, the State Space model can be expressed in many different functional forms under various different conditions, for example, continuous time-variant, continuous time-invariant, discrete time-variant, discrete time-invariant, Laplace domain of continuous time-invariant and Z-domain discrete time-variant. Meanwhile, there are different criteria for identification methods (Hamilton 1986).

If x_t is deterministic, the state vector ξ_t, concludes the all variables in the past which contribute for determining y_{t+m}.

$$E\left(y_{t+m} \mid \xi_t, \xi_{t-1}, \ldots, y_t, y_{t-1}, \ldots\right) = E\left(A'x_{t+m} + H'\xi_{t+m} + \omega_{t+m} \mid \xi_t, \xi_{t-1}, \ldots, y_t, y_{t-1}, \ldots\right)$$

$$= A'x_{t+m} + H'E\left(\xi_{t+m} \mid \xi_t, \xi_{t-1}, \ldots, y_t, y_{t-1}, \ldots\right)$$

$$= A'x_{t+m} + H'F^m\xi_t$$

Therefore, the State Space model can forecast the future value of y by observing x_{t+m} and ξ_t.

After discussing the State Space model, it is valuable to discuss the Kalman filter because it is usually used together with the State Space model. The Kalman filter is an algorithm in order to calculate $\left\{\hat{\xi}_{t+1|t}\right\}_{t=1}^{T}$ and $\left\{P_{t+1|t}\right\}_{t=1}^{T}$ while these are optimal forecasts of ξ_{t+1} under given observation from $(y_t, y_{t-1}, \ldots, y_1, x_t, x_{t-1}, \ldots, x_1)$ and mean squared error of the forecast respectively. Recalling the above equations:

$$\xi_{t+1} = F\xi_t + v_{t+1}$$

$$y_t = A'x_t + H'\xi_t + \omega_t$$

of which $E(v_{t+1}v'_{t+1}) = Q$ and $E(\omega_t \omega'_t) = R$. We can then assume that the parameters (F, Q, A, H, R) are certain. Based on this assumption, the optimal value in forecasting ξ_t on the basis of observation until time $t - 1$ can be calculated (Kalman 1960, Kalman 1963). Since the derivation of the Kalman filter is less essential in this section its application should be the major focus. Basically, it helps us forecast the stated vector at time $t + m$:

$$\hat{\xi}_{t+m|t} = E\left(\xi_{t+m} \mid y_t, y_{t-1}, \ldots, y_1, x_t, x_{t-1}, \ldots, x_1\right) = F^m \hat{\xi}_{t|t}$$

By subtracting the $\hat{\xi}_{t+m|t}$ from ξ_{t+m}, it can obtain the forecasting error by using the following equation:

$$\xi_{t+m} - \hat{\xi}_{t+m|t} = F^m\left(\xi_t - \hat{\xi}_{t|t}\right) + F^{m-1}v_{t+1} + \ldots + F'v_{t+m-1} + v_{t+m}$$

Meanwhile, the mean squared error will then become:

$$P_{t+m|t} = E\left[\left(\xi_{t+m} - \hat{\xi}_{t+m|t}\right)\left(\xi_{t+m} - \hat{\xi}_{t+m|t}\right)'\right]$$

$$= F^m P_{t|t}\left(F^m\right)' + F^{m-1}Q\left(F^{m-1}\right)' + F^{m-2}Q\left(F^{m-2}\right)' + \ldots + FQ'F + Q$$

By way of a similar process, the observed vector at time $t + m$, the equation can also be obtained as follows:

$$\hat{y}_{t+m|t} = E\left(y_{t+m} \mid y_t, y_{t-1}, \ldots, y_1\right) = A'x_{t+m} + H'F^m \hat{\xi}_{t|t}$$

The forecast error and mean squared error will then respectively be,

$$y_{t+m} - \hat{y}_{t+m|t} = H'\left(\xi_{t+m} - \hat{\xi}_{t+m|t}\right) + \omega_{t+m}$$

$$E\left[\left(y_{t+m} - \hat{y}_{t+m|t}\right)\left(y_{t+m} - \hat{y}_{t+m|t}\right)'\right] = H'P_{t+m|t}H + R$$

As a result, this process forecasts both the stated and observed values at time t + m. The forecast value at time t can be optimized according to the previous information until time $t - 1$ which is denoted as $\hat{\xi}_{t|t-1}$ and current available information at time t could be denoted as $\hat{\xi}_{t|t}$. If the available information at the end of the sample adjusts according to the historical value and the stated vector is taken in the middle of sample, it is called a smoothed estimate.

By using the Kalman filter, the sequences $\{P_{t|t}\}_{t=1}^{T}$ and $\{P_{t|t-1}\}_{t=1}^{T}$ can be obtained. Based on this result, it can further store the sequences $\{\hat{\xi}_{t|t}\}_{t=1}^{T}$ and $\{\hat{\xi}_{t|t-1}\}_{t=1}^{T}$. Thus, the terminal value of $\{\hat{\xi}_{t|t}\}_{t=1}^{T}$ can give a smoothed estimate at time T, denoted as $\hat{\xi}_{T|T}$ with $R_{T|T}$ as a mean squared error. Thus, through reversing order:

$$\hat{\xi}_{t|T} = \hat{\xi}_{t|t} + J_t\left(\hat{\xi}_{t+1|T} - \hat{\xi}_{t+1|t}\right), \text{ for } t = T-1, T-2, \dots, 1$$

where $J_t = P_{t|t}F'P_{t+1|t}^{-1}$ and the mean squared error will be:

$$P_{t|T} = P_{t|t} + J_t\left(P_{t+1|T} - P_{t+1|t}\right)J_t', \text{ for } t = T-1, T-2, \dots, 1$$

Although it seems to be quite complicated to apply the State Space model as an algorithm and it is hard to understand the concept of the model, the application is actually not as difficult as we expected. Various econometrics software, such as EView, Stata and R allow us to perform the State Space model easily.

2.6.1 The origins and advantages of the State Space model

The State Space model originated in the space program for tracking satellites. Since many of the computing systems could not generally provide enough memory, it motivated researchers to develop the recursive methods to perform forecasts and predictions. Thus, as in its original purpose, the state refers to the actual position of the satellites while the observation vector contains observed estimates of the satellite's location. Kalman (1960) discovered "state space" which aimed to solve the problems such as prediction of random signals, separation of random signals from random noise and detection of signals of known form in the presence of random noise. Wiener (1949) identified such declared problems with a solution method, spectral factorization, for the Wiener–Hopf integral equation. All these aimed to obtain the prediction, separation and detection of a random signal in a linear dynamic system until Kalman (1960) focused on "state."

Apparently, the previous methods had limitations that weakened their practical usefulness. First, there is an optimal filter which is specified by its own impulse response. However, it is too complicated to synthesize the filter. Second, it is difficult to compute the numerical determination optimal impulse response.

In fact, in recent times, such difficulty can be greatly decreased by technological improvements. Third, it is too specific, demanding strong techniques from researchers. Finally, the derivations consist of many assumptions that lead to the consequences becoming obscured.

Therefore, Kalman (1960) suggested that the State Space model had characteristics as follows. First, it provides optimal estimates and orthogonal projections. Second, State Space models concern random processes. The linear system can be specified by the system of I(1) equations. Third, by using the state transition method, it provides a solution for the Wiener problem (Wiener–Hopf integral equation). Furthermore, it can solve the dual problem of the noise-free optimal regulator. Finally, it is more practical than the previous methods.

2.6.2 Example 1: a study of the Guangzhou housing price bubble based on the State Space model

Hui and Gu (2009) made good use of a State Space model to measure the price bubble in the housing market in Guangzhou. It has gained popularity in the study of housing bubbles (for example, Teng *et al.*, 2013 and Hui *et al.* 2012). It should be noted that, even though the application of the State Space model is the same, its results would depend on the objectives of the researchers. In general, it is a popular tool to forecast prices and measure the housing bubble phenomenon.

Similarly to many cities in mainland China, Guangzhou has recently experienced high economic growth. Housing prices increased by 131.5% from 2004 to 2007. Although the financial crisis which originated in the US in 2008 affected exports and its housing market, housing prices were so high that many local residents could not afford to buy one property for their own use. The study used 60-months data from the Guangzhou Municipal Bureau of Land Resources and Housing Management and the Statistical Yearbook of Guangzhou in 2008. It covered the time period from January 2004 to December 2008 including housing price statistics and income per household,. The overview in price to income ratio suggested that there was a peak in late 2007.

The next step is to establish the State Space model. In general, the State Equation can be represented as follows:

$$\xi_{t+1} = F\xi_t + \upsilon_{t+1}, \, b_t = \varphi^* b_{t-1} + \varpi_t$$

where the observation equations can then be expressed as the following equations:

$$y_t = A'x_t + H'\xi_t + \omega_t, \, p_t = c_1 \, pq_t^* + b_t + \upsilon_t$$

where p_t, pq_t^*, q_t, b_t denote the housing price, price to income ratio, income and bubble.

Although there are a number of different definitions with regards to what constitute a bubble, the bubble term in State Space analysis usually refers to any unobservable state variable. The results showed that housing price and income per household are positively correlated such that the change in income per household would affect the growth of housing price. It was shown that the bubble was statistically significantly different from zero at the 5% level. This implied that there were some bubbles in the sample period. It concluded that housing bubble was highly correlated to the housing price fluctuation. However, the results of the test proved the existence of a bubble and it highly depended on researcher preference and the assumptions made.

2.7 The dichotomous Probit model

While many economic variables – GDP, energy consumption, land prices – are continuous variables, there is another group of variables which are dichotomous in nature. In this condition, the changes in independent variables lead to only two possible results, represented by one or zero (Vogt and Johnson 2011): for example, the success or failure of planning applications (Chau and Lai 2004), a collective decision to redevelop (Lai *et al.* 2012) housing with or without fittings (Li and Chau 2010). Under these cases, the estimation of the factors' impact on these zero or one dependent variables according to the traditional multiple regression will lead to error. Hence, the Probit and Logit models were invented for dichotomous dependent variables.

The Probit model provides a range of applications in econometrics and housing market analysis as well. For example, it is possible to construct a model to evaluate whether the applications have a high probability to receive a mortgage loan from a bank or end up with a mortgage default. Similarly, it is possible to set up a model for evaluating whether a person would stay or leave some regions' housing with regards to one particular school district in the city. Alternatively, it can be used to study the leading and lagging indicators in various housing price models.

The Probit model has a long history in application with an initiation study in Biology (Trevan 1927, Bliss 1935, Burn 1930). After that, many psychologists adopted the Probit model for mental tests (Ferguson 1942, Finney 1944, Lawley 1943). Some economists adopted it to study automobiles' property rights (Farell 1854). Until the late 1970s, the Probit model was adopted in urban economics with regards to home-ownership analysis (Bourassa 1995, Goodman 1988, Horioka 1988, Lee and Trost 1978, Painter 2000) and residential construction (Chan 1999). The linear probability model is often used to test the impact of the input of variables on the binary output (Cox 1970). The original form of the Linear Probability Model is:

$$E(Y_i|X_i) = Pr(Y_i = 1|X_i) = \beta X_i$$

where Y_i denotes the binary outcome which follows the Bernoulli probability distribution and X_i denotes the independent variable in continuous term. Consider if $Y_i = 1$ holds the probability, P_i, therefore, $1 - P_i$ will be the probability of $Y_i = 0$.

The Linear Probability Model holds the assumptions that the models are linear, exogenous and free from the problem of multi-collinearity. Thereby, the estimator β is not unbiased. Although it serves as the unbiased estimator, there are some critical drawbacks that greatly decreased its explanatory power. Even though there is no serial correlation, it is obvious that the disturbance is heteroscedastic because of the Bernoulli probability distribution. Meanwhile, the measurement of goodness of fit, R_2, is not appropriate in the linear probability model as well. The most serious defect is that it can compute a predicted probability whose value does not fulfill the condition:

$$0 \le \mathrm{E}(Y_i|X_i) \le 1$$

In order to solve the nonsense predicted probability feature, the Probit and Logit models were introduced. They use the concept of cumulative probability distribution function, to ensure that there will not be any nonsense predicted probability. As a matter of fact, both the Probit and Logit models capture non-linear populations well. They can produce similar results so that the selection between these two models is highly dependent on the preference of researcher. The difference between these two models is that the Logit model has fatter tails in a logistic distribution, therefore the change in outcome would be slightly slower than for the Probit model. In fact, the Logit model is slightly more preferred because of its mathematical simplicity. Nevertheless, this chapter will focus more on the Probit model as it has many further modifications by later contributors such as James Tobin (Gujarati and Porter 2009).

The Probit model uses a normal cumulative distribution function (CDF) that assumes the independent variables follow the normal distribution with a mean of zero and variance σ^2. The Probit model is established according to the following equations (Gujarati and Porter 2009):

$$\Pr(Y = 1|X) = \Phi(\beta X_i)$$

Given that it assumes normality, the probability that $Y_i^* \le Y_i$ can be obtained from the standardized normal CDF,[4]

$$P_i = \mathrm{P}(Y = 1|X) = P(Y_i^* \le Y_i) = P(Z_i \le \beta X_i) = F(\beta X_i)$$

where $P(Y = 1|X)$ denotes the probability that given X deduces $Y = 1$. Meanwhile, Z_i is the standard normal variable noted that $Z \sim N(0, \sigma^2)$. F refers to the standard normal CDF that

$$F(Y_i) = \frac{1}{\sqrt{2\pi}} \int_{-\infty}^{Y_i} e^{-z^2/2} \, dz$$

It follows the standard statistical table.[5]

---

4 "Standardized" means normal distribution with mean zero and unit variance.

5 The standard statistical table is provided in the appendix.

2.7.1 An application for the dichotomous Probit model

Consider the mortgage application with a special focus on the debt payment of applicants, a dichotomous Probit model can then be established to estimate the probability of unsuccessful application.

The Probit Regression model is designed with regressor X. The basic linear functional form has been used for illustration:

$$\Pr(Y = 1|X) = \Phi(\beta_0 + \beta_1 X_i)$$

where $Y = 1$ denotes the application is in futile and X_i is a variable which reflects the debt to income ratio of the applicant. After computing the regression, the Probit model is shown as follows:

$$\Pr(Y = 1|X) = \Phi(-2+3X_i)$$

At this moment, it is meaningless to explain the estimators in the Probit model. However, if it is in the linear probability model, there can probably be some meaning in the estimator since and it may carry the marginal effect. Thus, the further operation of the Probit model is to substitute a specific value of X_i in order to estimate the probability of success in a mortgage application.

For example, suppose there is an applicant who uses 40% of his income to pay for the debt, $X_i = 0.4$. By using the Probit model, it can be found that:

$$\Pr(Y = 1|X) = \Phi(-2 + 3 \times 0.4)$$

$$= \Phi(-0.8)$$

$$P(Z \le -0.8) = 1 - 0.7881 = 21.2\%$$

Finally, the probability of the application being denied is 21.2%.

2.7.2 Example 2: the formation of owners' corporations in Hong Kong's private housing estates: a Probit evaluation of Mancur Olson's group theory

In Hong Kong there is a legislative requirement in property management according to the Building Management (Amendment) Ordinance (Cap. 344). Many of the housing owners have formed their owners' corporations. The study collected 7,718 buildings from the Rating and Valuation Department in the Hong Kong Government. There were 100 buildings sampled from the analysis. The Probit model established:

$$OC = \Phi(c + b_1 \ln unit + b_2 urban + b_3 \ln age)$$

where Olson's Group $(OC) = 1$ denotes the presence of an owner corporation, and *urban* is another binary variable that classified the properties which were located in an urban district (=1) or otherwise 0. Moreover, *unit* and *age* denote the number of units in a residential building (or estate) and the age of the building (or estate) (Lai and Chan 2004).

The estimation results are statistically significant at 1% level which means they are highly significant. The results showed that the number of units has a negative relationship with owners corporation. Meanwhile, the location of a building had a positive relationship with the owners corporation. This further indicated that buildings in an urban district, it was estimated that there would be a higher chance to form an owners corporation than residential developments in the New Territories. Finally, the age of a building also exhibited a positive relationship with owners corporation. It indicated that older buildings or residential developments would be more likely to form an owners corporation. It showed that an older building needs to repair and renovate so the owners tended to form a management body rather than repairing by individuals (Lai and Chan 2004).

2.8 Conclusion

All in all, different econometrical models had been adopted to analyze the housing markets in different parts of the world. The Vector Error Correction model (VECM) and impulse response function were used to estimate the impact of an external shock such as financial crisis. They can be used in market fluctuation analysis especially for the stocks. They are then used to test whether the market can completely absorb a price shock before the next shock comes. Hence, the results are of practical and academic value. The Hedonic Pricing model estimates the incorporated demand and supply effect on housing price, especially for heterogeneous housing stocks and heterogeneous consumers, the environment, internal design and neighborhood. The model could be used to study the impact of housing construction and structures, external features, environmental pollution, facilities and location on the values of properties. The Cobb–Douglas production functions are applied to analyze various production factors. The State Space model can be used to solve the housing bubble phenomenon. Lastly, the Probit model offers good estimations when there is a binary dependent variable. All the models have their own specific characteristics and are applied in different housing markets analysis and scenarios. A thorough understanding of these econometric techniques is important for housing researchers.

2.9 References

Adalid, R. and C. Detken. (2007) Liquidity Shocks and Asset Price Boom/Bust Cycles. In *European Central Bank Working Paper Series*, 52. European Central Bank.

Ahumada, H. and M. L. Garegnani (1999) Hodrick–Prescott Filter in Practice. *Economica.-La Plata,* 45, 61–76.

Alessi, L. and C. Detken. 2009. Real Time Early Warning Indicators for Costly Asset Price Boom/Bust Cycles: A Role for Global Liquidity. In *European Central Bank Working Paper Series*, 56. European Central Bank.

Baxter, M. and R. G. King (1999) Measuring Business Cycles: Approximate Band-Pass Filters for Economic Time Series. *Review of Economics and Statistics,* 81, 575–593.

Berman, E. (1964) Resource Allocation in a PERT Network under Continuous Activity Time-Cost Function. *Management Science,* 10, 734–745.

Biddle, J. (2012) Retrospectives: The Introduction of the Cobb–Douglas Regression. *The Journal of Economic Perspectives*, 26, 223–236.

Bisping, T. O. and H. Patron (2008) Residential Investment and Business Cycles in an Open Economy: A Generalized Impulse Response Approach. *Journal of Real Estate Finance and Economics,* 37, 33–49.

Bliss, C. I. (1935) The Calculation of the Dosage–Mortality Curve. *Annals of Applied Biology,* 22, 134–167.

Borgy, V., L. Clerc and J. P. Renne (2009) Asset-Price Boom–Bust Cycles and Credit: What Is the Scope of Macro-Prudential Regulation? In *Banque de France Working Papers*, 1–62.

Borooah, V. K., J. Mangan and J. Hodges (1998) Determinants of Workplace Injuries: An Econometric Analysis Based on Injuries Compensation Data for Queensland. *Economic Analysis and Policy*, 28, 149–168.

Bourassa, S. C. (1995) A Housing Model of Tenure Choice in Australia. *Journal of Urban Economics,* 37, 161–175.

Burn, J. H. (1930) The Error of Biological Essay. *Physiological Review,* 10, 146–169.

Cebula, R. J. (2009) The Hedonic Pricing Model Applied to the Housing Market of the City of Savannah and its Savannah Historic Landmark District. *The Review of Regional Studies,* 39, 9–22.

Cecchetti, S. G. (2006) Measuring the Macroeconomic Risks Posed by Asset Price Booms. In *NBER Working Paper*, 29. The National Bureau of Economic Research.

Chan, T. S. F. (1999) Residential Construction and Credit Market Imperfection. *Journal of Real Estate Finance and Economics,* 18, 15.

Chao, W. (2003) Energy, the Stock Market, and the Putty-Clay Investment Model. *The American Economic Review*, 93, 311–323.

Chau, K. W. and L. W. C. Lai (2004) Planned Conversion of Rural Land: A Case Study of Planning Applications for Housing and Open Storage Uses in Agriculture Zones. *Environment and Planning B*, 31, 863–878.

Chau, K. W. and A. Walker (1988) The Measurement of Total Factor Productivity of the Hong Kong Construction Industry. *Construction Management and Economics*, 6(3), 209–224.

Chau, K. W. and A. Walker (1994) An Analysis of Factors Determining Project Level Labour Productivity of Construction Projects in Hong Kong. In *Proceedings of the CIOB International Conference on the Changing Roles of Contractors in Asia Pacific Rim*, 133–144.

Chau, K. W., S. K. Wong, A. T. Chan and K. Lam (2011) The Value of Clean Air in High-Density Urban Areas. In *High-Rise Living in Asian Cities*, 113–128. Springer, Netherlands.

Chau, K. W., S. K. Wong and C. Y. Yiu (2005) Improving the Environment with an Initial Government Subsidy. *Habitat International*, 29(3), 559–569.

Chau, K. W., C. Y. Yiu, S. K. Wong and L. W. C. Lai (2003) Hedonic Price Modeling of Environmental Attributes: A Review of the Literature and a Hong Kong Case Study. *Understanding and implementing sustainable development*, 87–110.

Chau, K. W. and R. D. Zhu (2009) The Characteristics of the Production Structure and the Nature of Technological Progress in Hong Kong's Construction Industry. *International Journal of Construction Management*, 9(1), 27–41.

Chin, T. L. and K. W. Chau (2003) A Critical Review of Literature on the Hedonic Price Model. *International Journal for Housing Science and Its Applications*, 27(2), 145–165.

Christensen, L. R., D. W. Jorgenson and L. J. Lau (1973) Transcedental Logarithmic Production Frontiers. *The Review of Economics and Statistics*, 55, 28–45.

Christiano, L. J. and T. J. Fitzgerald (2003) The Band-Pass Filter. *International Economic Review*, 44, 435–465.

Cobb, C. W. and P. H. Douglass (1928) A Theory of Production. *American Economic Review*, 18, 139–165.

Cogley, T. and J. M. Nason (1995) Effects of the Hodrick–Prescott Filter on Trend and Difference Stationary Time Series: Implications for Business Cycle Research. *Journal of Economic Dynamics and Control*, 19, 253–278.

Coulson, N. E. and M. Kim (2000) Residential Investment, Non-Residential Investment and GDP. *Real Estate Economics*, 28, 233–247.

Cox, D. R. (1970) Simple regression. In *Analysis of Binary Data*, 10. London: Methuen.

Dennis, E., B. Gordon and H. Sieg (2010) A New Approach to Estimating the Production Function for Housing. *American Economic Review*, 100, 905–924.

Detken, C. and F. Smets (2004) Asset Price Booms and Monetary Policy. In *European Central Bank Working Paper Series*, 62. European Central Bank.

Ekeland, I., J. J. Heckman and L. Nesheim (2002) Identifying Hedonic Models. *The American Economic Review*, 92, 304–309.

Engle, R. F. and C. W. J. Granger (1987) Co-Integration and Error Correction: Representation, Estimation, and Testing. *Econometrica*, 55, 251–276.

Farell, M. J. (1854) The Demand for Motor Cars in the United States. *Journal of the Royal Statistical Association A*, 117, 113–129.

Feng, C., L. Liu and S. Burns (1997) Using Genetic Algorithms to Solve Construction Time–Cost Trade-Off Problems. *Journal of Computing in Civil Engineering*, 11, 184–189.

Ferguson, G. A. (1942) Item Selection by the Constant Process. *Psychometrika*, 7, 19–29.

Finney, D. J. (1944) The Application of the Probit Analysis to the Results of Mental Tests. *Psychometrika*, 9, 31–39.

French, M. W. (2001) Estimating Changes in Trend Growth of Total Factor Productivity: Kalman and HP Filters Versus a Markov-Switching Framework. *FEDS Working Paper*.

Gattini, L. and P. Hiebert (2010) Forecasting and Assessing Euro Area House Prices Through the Lens of Key Fundamentals. In *European Central Bank Working Paper Series*, 48. European Central Bank.

Goodhart, C. and B. Hofmann (2008) House Prices, Money, Credit and the Macroeconomy. *Oxford Review of Economic Policy*, 24, 180–205.

Goodman, A. C. (1988) An Econometric Model of Housing Price, Permanent Income, Tenure Choice, and Housing Demand. *Journal of Urban Economics*, 23, 327–353.

Goodman, A. C. and T. G. Thibodeau (2003) Housing Market Segmentation and Hedonic Prediction Accuracy. *Journal of Housing Economics*, 12, 181–201.

Goodman, J. (1993) A Housing Market Matching Model of the Seasonality in Geographic Mobility. *Journal of Real Estate Research*, 8, 117–138.

Greiber, C. and R. Setzer (2007) Money and Housing: Evidence for the Euro Area and the US. In *Discussion Paper Series 1: Economic Studies, Deutsche Bundesbank*, 28. Deutsche Bundesbank.

Gujarati, D. N. and D. C. Porter (2009) *Basic Econometrics*. McGraw-Hill.

Hamilton, J. D. (1986) State-Space Models. In *Handbook of Econometrics*, eds. R. F. Engle and D. McFadden, Elsevier, 42.

Hassani, A. (2012) Applications of Cobb–Douglas Production in Construction Time–Cost Analysis. *Construction Systems*, 91. University of Nebraska, Lincoln.

Hodrick, R. J. and E. C. Prescott (1981) Postwar US Business Cycles: An Empirical Investigation. *Journal of Money, Credit and Banking*, 29, 1–16.

Horioka, Y. (1988) Tenure Choice and Housing Demand in Japan. *Journal of Urban Economics*, 24, 21.

Hui, E. C. M. and Q. Gu (2009) Study of Guangzhou House Price Bubble Based on the State Space Model. *International Journal of Strategic Property Management*, 13, 287–298.

Hui, E. C. M., C. Liang, Z. Wang, B. T. Song and Q. Gu (2012) Real Estate Bubbles in China: A Tale of Two Cities. *Construction Management and Economics*, 30(11), 951–961.

Intriligator, M. D., R. Bodkin and C. Hsiao (1996) *Econometric Models, Techniques, and Application*. Prentice Hall.

Kalman, R. E. (1960) A New Approach to Linear Filtering and Prediction Problems. *Transactions of the ASME–Journal of Basic Engineering*, 92, 35–45.

Kalman, R. E. (1963) New Methods in Wiener Filtering Theory. In *Proceedings of the First Symposium of Engineering Applications of Random Function Theory and Probability*, eds. J. L. Bogdanoff and F. Kozin, 119. New York: John Wiley and Sons, Inc.

Kaplanski, G. and H. Levy (2012) Real Estate Prices: An International Study of Seasonality's Sentiment Effect. *Journal of Empirical Finance*, 19, 123–146.

King, R. G., C. I. Plosser and J. H. Stock (1991) Stochastic Trends and Economic Fluctuations. *American Economic Review*, 81, 819–840.

King, R. G. and S. T. Rebelo (1993) Low Frequency Filtering and Real Business Cycles. *Journal of Economic Dynamics and Control*, 17, 207–231.

Koop, G., M. H. Pesaran and S. M. Potter (1996) Impulse Response Analysis in Nonlinear Multivariate Models. *Journal of Econometrics*, 74, 119–147.

Lai, L. W. C. and P. Y. L. Chan (2004) The Formation of Owners' Corporations in Hong Kong's Private Housing Estates: A Probit Evaluation of Mancur Olson's Group Theory. *Property Management*, 22, 55–68.

Lai, L. W. C., K. W. Chau and J. W. Y. Kwong (2012). Surrendering the Environment for Capital Gain and Olson's Group Theory: A Case Study of the Dissolution of Housing Co-operative Societies in Hong Kong. *Habitat International*, 36(2), 261–267.

Lamberson, L. and R. Hocking (1970) Optimum Time Compression in Project Scheduling. *Management Science*, 16, 597–606.

Laurice, J. and R. Bhattacharya (2005) Prediction Performance of a Hedonic Pricing Model for Housing. *The Apprasial Journal*, 73, 198–209.

Lawley, D. N. (1943) On the Problems Connected with Item Selection and Test Construction. *Proceedings of the Royal Society of Edinburgh*, 61, 15.

Lee, L. F. and R. P. Trost (1978) Estimation of Some Limited Dependent Variable Models with Application to Housing Demand. *Journal of Econometrics*, 8, 357–382.

Li, R. Y. M. (2009) The Impact of Climate Change on Residential Transactions in Hong Kong. *The Built and Human Environment Review*, 2, 11–22.

Li, R. Y. M. and K. W. Chau (2010) An Empirical Study of Brand Name, Land Costs, Income and Housing Availability's Impact on Residential Fittings in Hanzhou, China. In 16th Annual *Pacific Rim Real Estate Society Conference*, Wellington, New Zealand

Li, R. Y. M. and R. Hung (2013) Rostow's Stages of Growth Model, "Urban Bias" and Sustainable Development in India. *Journal of Contemporary Issues in Business Research*, 2, 170–178.

Li, R. Y. M. and C. Y. Ng. (2013) The Chicken-And-Egg Relationship Between Foreign Direct Investment Stock and Economic Growth in South Africa. *Journal of Current Issues in Finance, Business and Economics*, 6, 23–38.

Li, R. Y. M., C. Y. Law and T. H. Leung. (2014) Hong Kong People are No Longer Superstitious? The Pricing of Residential Units' Luckiness Revisited. In *19th Asian Real Estate Society,* Brisbane, Australia.

Mcgough, T. and S. Tsolacos (1995) Property Cycle in the UK: An Empirical Investigation of the Stylized Facts. *Journal of Property Finance* 6, 45–62.

McGreal, S. and P. T. D. L. Paz (2013) Implicit House Prices: Variation Over Time and Space in Spain. *Urban Studies,* 50, 2024–2043.

Malpezzi, S. (2002) Hedonic Pricing Models: A Selective and Applied Review. In *Housing Economics and Public Policy,* eds. T. O'Sullivan and K. Gibb. Wiley.

Miller, N. G., V. Sab, M. Sklarz and S. Pampulov (2013) Is There Seasonality in Home Prices? Evidence from CBSAs. *Journal of Housing Research,* 22, 1–15.

Moussourakis, J. and C. Haksever (2010) Project Compression with Nonlinear Cost Functions. *Journal of Construction Engineering and Management,* 136, 251–259.

Ng, S. and Y. Zhang (2008) Optimizing Construction Time and Cost Using Ant Colony Optimization Approach. *Journal of Construction Engineering and Management,* 134, 721–728.

Painter, G. (2000) Tenure Choice with Sample Selection: Differences Among Alternative Sample. *Journal of Housing Economics,* 9, 197–213.

Park, G. (1996) The Role of Detrending Methods in a Model of Real Business Cycles. *Journal of Macroeconomics,* 18, 479–502.

Pesaran, M. H. and Y. Shin (1998) Generalized Impulse Response Analysis in Linear Multivariate Models. *Economics Letters,* 58, 17–29.

Ravn, M. O. and H. Uhlig (2002) On Adjusting the Hodrick–Prescott Filter for the Frequency of Observations. *The Reviews of Economics and Statistics,* 84, 371–375.

Ronayne, D. (2011) Which Impulse Response Function? In *Warwick Economic Research Papers*. Department of Economics, The University of Warwick.

Rosen, S. (1974) Hedonic Prices and Implicit Markets: Product Differentiation in Pure Competition. *Journal of Political Economy,* 82, 34–55.

Rowell, D. (2002) *State Space Representation of LTI Systems*. Universidad de Michigan.

Rush, R. and T. H. Bruggink (2000) The Value of Ocean Proximity on Barrier Island Houses. *The Apprasial Journal,* 9.

Shen, Z., R. Shamsi and A. Hassani. (2012) A New Perspective for Construction Crashing Cost Analysis. In *Proceedings of International Conference on Construction and Real Estate Management*. Kansas City, US.

Sirmans, G. S., D. A. Macpherson and E. N. Zietz (2005) The Composition of Hedonic Pricing Models. *Journal of Real Estate Literature,* 13, 1–14.

Stock, J. H. and M. W. Watson (1999) Forecasting Inflation. *Journal of Monetary Economics,* 44, 293–335.

Tekel, A. and L. Akberishahabi (2013) Determination of Open-Green Space's Effect on Around House Prices by Means of Hedonic Price Model: In Example of Ankara Botanik Park. *Gazi University Journal of Science,* 26, 347–360.

Teng, H. J., C. O. Chang, and K. W. Chau (2013) Housing Bubbles: A Tale of Two Cities. *Habitat International*, 39, 8–15.

Trevan, J. W. (1927) The Error of Determination of Toxicity. *Proceedings of the Royal Statistical Association,* 101, 483–514.

Vogt, W. P. and R. B. Johnson. (2011) *Dictionary of Statistics and Methodology: A Non-Technical Guide for the Social Sciences*. Sage, Thousand Oakes, CA.

Wiener, N. (1949) *Extrapolation, Interpolation and Smoothing of Stationary Time Series: With Engineering Applications*. The M.I.T. Press.

Witkiewicz, W. (2002) The Use of the HP Filter in Constructing Real Estate Indicators. *The Journal of Real Estate Research,* 23, 65–88.

Wolverton, M. L. and S. C. Bottemiller (2003) Further Analysis of Transmission Line Impact on Residential Property Values. *The Apprasial Journal*, 9.

Xiao, Q. and R. G. K. Tan (2007) Signal Extraction with Kalman Filter: A Study of the Hong Kong Property Price Bubbles. *Urban Studies,* 44, 865–888.

3 Risk averse real estate entrepreneurs in mainland China

A Probit model approach

Rita Yi Man Li and Kwong Wing Chau

3.1 Introduction

In many places around the world, such as Australia, New Zealand, Hong Kong, the United Kingdom and the United States, housing developers provide floor and wall coverings, window frames, cupboards, and electrical fittings as standard equipment and fittings. In some cases, real estate developers also provide heated floors and wine storage for homebuyers, i.e. the fittings are tied in together with the housing units no matter whether the homebuyers like it or not. In any case, however, such arrangements provide a level of convenience to homebuyers if they do not have a special preference on the design of fittings. Some advertisements even suggest that homebuyers can simply carry their luggage to move in.

There is, however, a different norm in mainland China. For example, the majority of real estate developers in Beijing and Nanjing sell bare dwellings to homebuyers. There are no floor coverings, kitchen cupboards or basic bathroom fittings (Li 2009a). Some said the fittings provided by the developers were poor (Li and Chau 2014). Alternatively, we may view this as the result of informal institutions such as culture, and formal institutions such as legal systems (Li 2010a, 2010b, Li 2011b).

Among all the residential developments available for sale from 2004 to 2007 in Nanjing, 90% were bare units. Beijing displays a similar phenomenon from 1997 to 2008. Of all the first hand residential units sold in Beijing districts, more than 70% were sold as bare units. Similarly, bare units can also be found in Shanghai and Hangzhou. This chapter aims at finding out the major drivers behind these housing entrepreneurs to build bare residential units in these cities.

3.2 Four hypotheses

Purchasing bare units without any fittings in mainland China implies that homebuyers need to spend time in identifying relevant parties to decorate their housing units before they move in. Why do busy people, often trapped in crammed schedules accept such a time consuming building activity? There is very limited study

Table 3.1 Percentage of bare residential units in Nanjing (Li 2010b)

Districts Nanjing	Total residential developments available for occupation in 2004–2007*	Percentage of bare flats
Bai Xia	32	92
Da Han	4	50
Gao Chun	3	100
Gu Lou	65	94
Jian	71	93
Jiang Zhu	133	98
Li Shui	11	91
Lu He	12	92
Pu Kou	60	97
Qi Xia	50	100
Qian Huai	42	100
Xia Guan	24	92
Xuan Wu	35	92

* Each of the residential developments refers to one residential project. Some of the projects consist of up to, or more than 1,000 units. The districts refer to small areas inside the city.

Table 3.2 Percentage of bare flats in Beijing from 1997 to 2008 (Li 2009a)

Districts in Beijing	Residential projects which build bare flats	Total	Percentage of bare flats
Chao Yong	618	853	72
Chong Ping	160	182	88
Chong Wen	62	74	84
Da Xing	140	165	85
Dong Cheng	59	79	75
Fang Shan	83	90	92
Feng Toi	260	305	85
Hai Dian	379	470	81
Huai Rou	37	44	84
Mi Yun	43	52	83
Shi Jing Shan	48	54	89
Shun Yi	102	126	81
Tong Zhou	194	219	89
Xi Cheng	62	82	76
Xuan Wu	88	114	77
Yan Qing	17	24	71
Others	92	110	84

of this phenomenon. This chapter aims to remedy this and test the potential reasons behind this by testing theories of brand name, risk averse behavior, information costs, etc. Specifically, we propose four hypotheses on the emergence of bare flats:

1   Developers tend to build housing units with fittings in high land price areas.
2   The larger the proportion of low income residents the higher the proportion of bare flats.
3   Developers with well renowned reputations (brand names) build more furnished flats.
4   The greater the shortage of residential units, the higher the proportion of bare flats.

3.2.1 The first proposition: developers tend to build housing units with fittings in areas with high land prices

While fittings in high land price areas only make up a small proportion in construction costs, fittings in low land price areas constitute a relatively large share. The costs of providing fittings which do not suit the taste of customers are relatively lower in areas with high land prices compared to the costs of providing wrong fittings in areas with low land prices as home purchasers might choose not to buy the flats when developers provide unsuitable fittings. As developers are risk averse, it is natural that they provide fewer fittings in those areas to avoid the relatively high risks:

$$P(f)/L_H < P(f)/L_L$$

where $P(f)$ refers to the price of the fittings, L_H refers to high land price area and L_L refers to low land price area.

3.2.2 The second proposition: the larger the proportion of low income residents, the higher the proportion of bare flats

Developers in affluent countries – such as, the United Kingdom, Singapore, the United States – offer more fittings as compared to those in developing countries, such as China, Ghana, Indonesia etc. Buying favorite fittings implies that the buyers are investing the time and money in searching for information. A man who earns $4 per hour has a low discretionary income but may be happy to search by himself rather than pay the contractor to do all on behalf of him, while a man who earns $ 400 finds it not worthwhile to do so. Searching costs him $400 per hour which is more than the cost of letting the contractor do it. In view of this, high income areas theoretically should have a smaller portion of bare flats. Developers tend to build more well-furnished flats with fittings to suit the needs of customers.

3.2.3 The third proposition: developers with well-developed reputations build more well-furnished flats and vice versa

Bare dwellings may avoid losses but do not help the developers to maximize gains from the value added. Hence, well regarded developers may tend to build more housing units with good fittings to enhance their reputation for quality. For others, it may be better to provide no fittings as the provision of poor fittings may set back reputations which have taken years to build up.

3.2.4 The fourth proposition: the supply is so limited that the huge demand lowers developers' motivations to provide well-equipped units

Previous literature shows that shortages may lead to the use of unqualified staff and poor materials. Compared to many overseas countries, China experiences a shortage of private housing. With substantial excess demand, developers do not need any gimmicks to compete for potential buyers. Provisions of kitchen and bathroom fittings are unnecessary.

3.3 Literature review

3.3.1 Assumptions, functions and factors which affect the supply of entrepreneurs

Entrepreneurs have been seen as the engine which propel much of the growth of the business sector and economic growth (Noruzi *et al.* 2010, Fairoz *et al.* 2010). They run business, provide business strategies and future plan and hire workers. For example, housing companies' entrepreneurs hire architects, surveyors, engineers, computer programmers and so on. Another important role played by entrepreneurs is risk-taking. Entrepreneurs are risk-takers (Blanchflower and Oswald 1998) who accept firms' risks (Carland *et al.* 1984) when they run a business. In times of good economy, many individuals start up their own companies and become entrepreneurs. Nevertheless, when there are financial crises or the economy is bad, many entrepreneurs close their business or are forced to leave due to bankruptcy. Housing developers are classified as entrepreneurs who own licenses that allow them to construct houses for homebuyers and generate reasonable profits from their investments (Jaafar *et al.* 2014).

3.3.2 Imperfect information and asymmetric information

In classical economics, information is assumed to be perfect. Transitory assemblages based on intentional cooperative agreements could govern utility-driven production (Valentinov 2008). Each of us has an equal set of information, which can be obtained at zero cost. The later New Institutional economics, however, suggests that there are transaction costs and information can be costly (Li 2011b, Li 2014a, Li 2014b). We need to pay for information and we have to spend time and effort to obtain it.

The empirical evidence goes better with the later theoretical framework provided by the New Institutional Economists. Information is not perfect and asymmetric information in many occasions (Li 2014a, Li 2014b). For example, developers have more information about the quality of fittings than the homebuyers. Bankers have less information than the money borrowers with regards to bad debt. The farmer who rents a piece of farmland knows the soil fertility better than the landlord. In most circumstances, the farmer is the best "valuer" to value the monthly rent which he should pay (Li 2011c).

In the first hand residential units sale, one major problem that the entrepreneurs face is imperfect information. Imperfect information affects market power and prices which leads to a dis-equilibrium where firms charge monopoly prices (Granlund and Rudholm 2011). Because of the existence of information and transaction costs, different people have different information, i.e. asymmetric information exists (Li 2015). There is false information embodied in unenforceable contracts and worthless promises (Li 2014).

Asymmetric information is one example of a market failure (another is a monopoly) which undermines the efficiency of product transactions. It is not unusual that consumers are uninformed about risks; this affects their ability to choose terms which reflect their preferences correctly. Firms often exploit this ignorance by degrading contract quality intentionally and will then have little incentive to offer better deals as these will not increase sales (Bechern 2008). Previous research also suggests that asymmetric information exists between firm and government (Du *et al.* 2014), developers and contractors (Li 2009b).

3.3.3 Risk averse human behavior

Because of imperfect information, individuals and entrepreneurs make their decisions under risk. Risk has been identified as the potential for threat, damage, injury, or other loss (Zou *et al.* 2007, Jin and Doloi 2008). It may also be conceptualized as variance in the outcome of a project (Das and Teng 2001). Risk can be managed by catastrophe planning, easing, insurance, control, identification, quantification and shifting it to other agents (Zou *et al.* 2007, Jin and Doloi 2008).

People's attitude to risk affects their behavior and is important in decision making (Richard 1975, Xiao and Yang 2008). Ever since Bernoulli, research has shown that individuals are risk averse. We try to avoid risk by various means. In fact, risk aversion provides a vivid explanation for many social phenomena (Li and Poon 2011). Nevertheless, people's attitude towards it is not consistent. Some people are more risk averse than the others. Moreover, risk is not bad in every single aspect. It is considered that risk-taking is an essential component of a progressing and a progressive society (Basham and Luik 2011).

In the Tasmanian housing sector, minimizing the perceptions of risk associated with social housing tenants was seen as an important and necessary condition to promote investment and frame housing assets among residential managers (Francis-Brophy and Donoghue 2013). To reduce risk, landlords in Greece often use a discriminatory rating system to estimate the risks. For example, there is restriction on the supply of residential units available to Albanians in luxury areas (Drydakis 2011).

In first hand housing sale, buyers cannot obtain information on housing quality from the previous owners when they buy the new ones. Many of them only know they have bought poor housing when they open the doors of their units (Gwin and

Ong 2000). Therefore, homebuyers often rely on developers' reputation and previous residential projects to make decisions so as to minimise their risk in home purchases.

3.3.4 Loss-averse human behavior

Apart from risk aversion, many entrepreneurs are loss-averse, i.e. decision-makers are more sensitive to losses than to gains. This phenomenon represents a discontinuity in their utility function (Berkelaar, Kouwenberg and Post 2004), graphically expressed as an abrupt change in the slope of the utility function at the reference point (Wang *et al.* 2009) which distinguishes gains from losses. It also signifies that the utility function is steeper for losses than for gains, i.e. the disutility that one experiences in losing money is greater than the utility associated with gaining the same amount. Loss aversion has become an important tool to explain all sorts of phenomena which are not explained by traditional theory, such as the endowment effect (Tovar 2009).

In housing, prices fall below the homeowners' original purchase price set an asking price that exceeded the asking price of other sellers between 25 and 35%. Both investors and owner-occupants are loss averse. The evidence of loss aversion not only rests on asking prices or on sellers who could not sell their houses. Sensitivity of the asking price to nominal loss among successful sellers was only around 50% of homeowners who finally withdrew from the market. In times of uncertainty, the value function is steeper for losses than for similar gains. Furthermore, the marginal value of losses or gains diminishes with the size of the loss or gain. Put together, these attributes trace out the familiar value function from prospect theory (Genesove and Mayer 2001).

3.3.5 Economic value of branded product

A brand name is a firm's most important asset when it comes to evaluating companies against each another (Laforet 2011). It is a commodification process which makes things and people respected, idolized, adored, worshipped (Preece and Kerrigan 2015). Brand name affects the individual purchase decision: it forms a subconscious mental representation or memory which can affect the subsequent interpretation, evaluation and perception of the stimulus. For instance, mental representation of a brand name often eases acceptance of the product by providing a feeling of familiarity (Janiszewski 1993).

It is a useful tool for many businessmen to change their products from elastic to inelastic, i.e. more likely to buy their products even when prices increase. For example, when a similar design of handbags is on display, the LV handbag will sell at a higher price than the non-branded products. It is no wonder why many entrepreneurs spend a lot of effort on boosting the name of their companies. TV, train, Youtube and Facebook advertisements have become indispensable in the eyes of some large scale corporations. Superstars are hired to enhance their product's brand name. Indeed, branded products have valuable merits to consumers as well as sellers as they convey quality information to the consumer. As early as the 1960s, the American Marketing Association defined "brand" as:

> [A] name, term, sign symbol or design, or a combination of these, intended to identify the goods or services of one seller or group of sellers and to differentiate them from those of competitors.
>
> (Zilg 2011, p. 284)

Sellers associate a good brand with monetary benefits. Consumers often take a shorter time to make purchasing decisions when they buy more reputable products (Pan *et al.* 2015) as brand is an important clue to product quality. A reputable brand name often helps stabilize the quality perceptions of a product even when the price is lowered. Brand also positively affects purchasers' internal reference prices: previous experience provides an important stimulus. One component of past experience includes brand name recognition. Hence, even when consumers have no direct experience with a product, exposure to brand name provides a certain level of familiarity (Grewal *et al.* 1998).

In a lot of occasions, there is a positive relationship between internal reference price and brand name. Moreover, brand names affect buyers by affecting their internal reference prices via changes in their perceptions of brand or merchandise quality (Grewal *et al.* 1998). The total cost of psychopharmaceuticals was raised by 20.1% more for branded ones than the generic agents (Polić-Vižintin *et al.* 2014). It is no wonder that franchisees have an incentive to free ride on the brand name of the franchisor (Ricketts 2015).

In housing, brand is also important. Previous research shows that an appropriate naming strategy provides valuable information for consumers, which expedites homebuyers' decision making. Moreover, suitable names which are liked by a substantial number of consumers can increase the demand and push up the housing prices (Chau *et al.* 2007, Zahirovic-Herbert and Chatterjee 2012). Chau *et al.* (2001)'s research found that people are willing to pay a premium of at least 7% of the housing price for developers' good will on average. Hence, it was suggested that appraisers should consider the reputation of developers when they perform valuation duties. Even though developers' good will has not been used traditionally to attract consumers, we can consider it a profitable investment in the long run.

With asymmetric information in China's housing industry, buyers use brand names to assess product quality. Reputation is an effective signal which provides quality information because it is firm-specific – gradually built up from the quality of the projects built by a developer in the past. Homebuyers can obtain information on quality of units by observing previous projects.

3.3.6 Shortage and sellers' goods quality

Market forces not only affect prices, they also determine suppliers' incentives to provide "extra" value which aims at being perceived as quality improvement. In

Table 3.3 Merits of brand name

Merits of brand name	Examples
It allows firms to escape from the confines of generic prices.	Jensen and Drozdenko (2008); LeBel and Cooke (2008); Rotfeld (2004); Vukasovič (2009); Souiden and Pons (2009)
It becomes easier to adopt a new product.	Vukasovič (2009)
There is price premium for branded products.	Olson (2008); Chen (2007); Li *et al.* (2009) disagree)
It enhances perceptions of product or service quality	Horppu *et al.* (2008); Shannon and Mandhachitara (2008); Vukasovič (2009);
It facilitates promotional effectiveness.	
The "personality" of brand product adds value to the consumer.	
It increases market share.	Vukasovič (2009)
It decreases risk.	Vukasovič (2009); Matzler *et al.* (2008)
It decreases consumers' time to make purchase decisions	Pan *et al.* (2015)
Barrier to entry for competitors	Omar *et al.* (2009)
It provides information to the consumer.	Baltas and Saridakis (2009); Pechtl (2008); Pitta and Franzak (2008); Souiden and Pons (2009)
It enhances product differentiation	Zilg (2011)
It is important in commercialization and marketing.	Pires *et al.* (2015)
It increases prices and sales	Zahirovic-Herbert and Chatterjee (2012); Polić-Vižintin *et al.* (2014)
To attract a group of loyal customers.	Burnett and Bruce (2007); Horppu *et al.* (2008)

a shortage, suppliers do not need to compete with other sellers. As a result, they may minimize production costs by providing only limited accessories to their buyers (Hawthorne and Birrell 2002, Ingersoll and Smith 2003).

Housing investment in the planned economy was limited before 1976. The government preferred to spend money on investments other than housing (Wang and Murie 1996). Although the government no longer distributed housing to citizens and private housing investment was allowed after the death of Mao, rapid growth in population and subsequent alterations in national urban policy had led to a rapid urbanization in China. Registered urban population has been increasing at a rate of 4% annually since 1980 (Wu 1999). Residential unit supply never met the demand. Housing shortages problems had become serious by the end of the Cultural Revolution in 1976 when there was only 3 m² of floor space per person on average. Despite the large quantity of housing built since 1978, four million urban households still live with 4 m² per family member in 1994 and 0.4 million households occupied an average of 2.5 m m² living space per person (Wang and Murie 1996).

In view of the above, does it mean that the popularity of bare units in China is the result of a shortage of housing, risk averse behavior by developers or something else?

3.4 Research method

The majority of home developers worldwide tie in various fittings – washing basins, floor tiles, etc. – with the housing for sale. It will be interesting to study the various factors which lead to the sale of bare units. We will use the Probit model. In a binary response model, academic interest lies primary in the response probability in zero or one dependent variable. Nevertheless, the dichotomous distribution implies that the error term can also be dichotomous in nature; thus the normally distributed assumption may not be correct. Therefore, OLS regression – suitable for continuous dependent variables – may not be the best choice in case of dichotomous dependent variables to estimate an equation (Gujarati 2006).

Probit models have been adopted in much previous research. For example, Li (2011) applied it to study the factors which affected successful planning applications. Lai and Chan (2004)'s paper adopts a Probit model to study the likelihood to form owners' corporations in Hong Kong under the lens of Mancur Olson's group theory. It was found that older urban estates with fewer owners were more likely to form owners' corporations.

$$Yt^* = Xt$$

$$P(y = 1|x) = P(y1 \mid x_1, x_2, \ldots, x_k)$$

Consider an equation with binary response (0,1):

$$P(y = 1|x) = Y(\eth_0 + \eth_1 x_1 + \eth_2 x_2 + \ldots + \eth_k x_k) = Y(\eth_0 + x\, \eth)$$

where x denotes a complete set of explanatory variables, Y is a function which takes on values between zero and one, that is $0 < Y(z) < 1$ where t represents all the real numbers (Wooldridge 2003):

$$x\, \eth = \eth_1 b_1 + \eth_2 b_2 + \ldots + \eth_k b_k$$

3.5 Data analysis

To achieve our objectives, we have collected some of the macro economic data. Data of provincial GDP per capita, construction costs for each year, annual supply of newly built residential units and change in number of household, background information of the residential developments such as developers which built the dwellings were collected to test the four hypothesis listed at the beginning of the chapter.

Per capita data of 2003–2007 GDP for three places – deflated by the GDP deflator – were obtained from the Statistics Bureau in each place (Shanghai Statistics Bureau 2008, Chongqing 2008, Hangzhou Statistics Bureau 2008). Land price was a proxy obtained by subtracting the average selling price of residential units (Shanghai Soufun 2009, Wooldridge 2003, Soufun 2009) by construction costs available from the respective Statistical Year Books of (Shanghai Statistics Bureau 2008, Chongqing 2008, Hangzhou Statistics Bureau 2008).

In this five-year period, about 80% of the dwellings in Hangzhou were bare units. The number of developments ready for sale has been fairly constant at about 50–60 per year. The real GDP per capita[1] rose from 32,667 to 52,590 in 2003 and 2007 respectively. Construction costs have increased since 2004. Estimated land costs have risen from 9350 in 2003 to 10,484 in 2006 but dropped to 6616 in 2007. GDP per capita has increased from 32,667 in 2003 to 52,590 in 2007. Construction costs per m^2 and number of residential developments rose from 827 to 1452 and 47 to 151 during the five years' observation period respectively. In Shanghai, the percentage of bare units dropped from 88 to 80 between 2003 and 2007. Number of housing developments rose from 149 to 304 in 2003 and 2005 but decreased afterwards and reached a trough of 128 in 2007. Construction costs oscillated around 2900 in this five year period. Annual change in number of household rose from 42,900 to 61100 between 2003 and 2005, dropped to 28,500 in 2006 and rose to 37,500 in 2007. Area of newly built dwellings climbed from 22,810,000 m^2 in 2003 and reached a maximum at 32,700,000 m^2 in 2004.

3.6 Research results and analysis

To test the hypothesis, five variables LDMV, BUILDONCE, BUILDONCE* LOG(CCMV), A_CHANGEHSE, and GDP were included.

The Probit Regression models confirm the hypothesis that there is a signifi-cant negative relationship between bare units and land costs: there is a higher probability that developers build bare units when the land costs are low. Costs of fittings provisions become relatively high in low land costs areas as compared to the whole construction costs. On top of that, buyers may choose not to pur-chase units if the fittings do not suit their tastes. China is a newly emerging real estate market as there were no private companies prior to 1978 due to the planned economy. Hence many housing entrepreneurs are newcomers to the industry and lack information for predicting buyers' tastes in house fittings. In sharp contrast, the well-established markets in places, such as Hong Kong, Europe, and Australia have enabled local housing entrepreneurs in these locations to learn about the prefer-ences of buyers. Housing developers are risk averse. They all aim to maximize their

---

1 All costs and prices are expressed in RMB unless otherwise specified. Currently 1 USD equals 6.23 RMB.

Table 3.4 Macroeconomics, land and housing data in Hangzhou, Chongqing and Shanghai

Year	Real GDP per capita	Construction costs (m²)	Number of development	Average estimated land cost (per m²)	Change in no. of households (ten thousand)	Area of newly built dwelling ten thousand (m²)	Percentage of bare development (per cent)
Hangzhou (Hangzhou Statistics Bureau 2008, Hangzhou Soufun 2009)							
2003	32667	2275	26	9350	2.85	516	82
2004	38593	2023	61	11435	3.40	528	83
2005	44555	2459	62	11535	2.90	662	83
2006	51650	2340	51	10484	2.49	580	81
2007	52590	2953	59	6616	2.08	675	81
Chongqing (Chongqing 2008, Hangzhou Soufun 2009)							
2003	7261	827	47	3242	15.32	1232	85
2004	7633	929	80	3281	11.58	1228	90
2005	7745	1074	106	3694	21.82	1714	93
2006	7689	1319	146	2800	20.25	1700	93
2007	7854	1452	151	3023	26.31	1769	93
Shanghai (Shanghai Statistics Bureau 2008, Shanghai Soufun 2009)							
2003	39128	2987	149	6294	4.29	2 281	88
2004	46338	2958	180	8107	4.52	3 270	83
2005	51529	2960	304	7204	6.11	2 819	84
2006	57695	2963	151	8491	2.85	2 747	80
2007	66367	3014	128	9319	3.75	2 844	80

Table 3.5 Variables which are included in the Probit model

Variables	Explanation
LDMV	Land price divided by open market value of the dwellings that is, selling price of the units.
BUILDONCE	It represents developers which built once only during the study period. As branded developers are characterized by repeated sales, dummy "1" is assigned if the housing development was built by a developer which had built once only and "0" if they had engaged in repeated sales activities from 2004–2007.
A_CHANGEHSE	To test the impact of housing shortage on provision of fittings, dwelling space was divided by the change in number of household (Powell and Ansic 1997) and it is represented by A_CHANGEHSE.
BUILDONCE*LOG(CCMV)	CCMV denotes the proportion of construction cost in dwelling value.

Table 3.6 Results of Quadratic hill climbing Probit regression

Dependent Variable: BARE
Method: ML–Binary Probit (Quadratic hill climbing)
Included observations: 1704
Convergence achieved after five iterations
Covariance matrix computed using second derivatives

Variable	Coefficient	Std. error	z-Statistic	Prob.
C	2.320284	0.698542	3.321609	0.0009
LOG(LDMV)	−0.645794	0.233331	−2.767715	0.0056
BUILDONCE*LOG(CCMV)	0.614759	0.129308	4.754215	0.0000
BUILDONCE	0.919487	0.193384	4.754730	0.0000
LOG(A_CHANGEHSE)	−0.259066	0.100172	−2.586210	0.0097
LOG(GDP)	−0.007551	0.113477	−0.066540	0.9469
McFadden R-squared	0.106837	Mean dependent var		0.861502
S.D. dependent var	0.345523	S.E. of regression		0.325198
Akaike info criterion	0.725551	Sum squared resid		179.5704
Schwarz criterion	0.744708	Log likelihood		−612.1692
Hannan-Quinn criter.	0.732642	Restr. log likelihood		−685.3946
LR statistic	146.4507	Avg. log likelihood		−0.359254
Prob(LR statistic)	0.000000			
Obs with Dep = 0	236	Total obs		1704
Obs with Dep = 1	1468			

profits without too much risk. To reduce their risk, these entrepreneurs choose not to provide any fittings. Furthermore, traditional entrepreneurs rely on many sources of external financing: debt, venture capital, private equity and public stock offerings. Investors who commit funds to business start-ups expect to receive back their investment along with a handsome return. Entrepreneurs must ask themselves if their goals are congruent with those of possible investors (Williams *et al.* 2006).

If housing entrepreneurs make an incorrect decision, such as putting undesirable fittings into their units, investors will be affected and develop negative opinions about these entrepreneurs. Potential joint ventures may also become futile if the housing entrepreneurs have a history of incorrect decisions. Under the postulate of being risk averse, developers prefer to provide nothing inside the units.

The Probit Regression results also confirm our hypothesis that there is a significant positive relationship between repeated sales activities and furnished units. Nonetheless, the results of the interaction term BUILDONCE*LOG (CCMV) shows that not all "new" developers are equally keen to construct bare housing. The higher the construction cost as a proportion of property price, the higher would be the risk that the monetary return cannot cover the cost.

The literature indicated that shortages can lead to lower consumer requirement, and this is supported in this Probit model – there is an increase in likelihood of bare units construction in times of housing shortage. However, the insignificant relationship between GDP per capita and bare units, implies that people with higher information costs may not necessarily prefer well-furnished units.

3.7 Conclusion

An entrepreneur is the major driving force behind production of goods as he/she combines the resources of land, capital and labor. All entrepreneurs share one characteristic: they bear risk and earn profit or suffer losses in return. It is assumed here that housing entrepreneurs are risk averse. They attempt to find ways to reduce their risk. Housing entrepreneurs in mainland China who provide bare units provide a vivid example of risk aversion. As there is no guarantee of making a profit and there is imperfect information in our business world, the entrepreneur must assess the risk and, like other entrepreneurs, he/she cannot accept a job that has too high a risk of loss (Brue *et al.* 2009).

Running a real estate development business is risky; it requires substantial capital investment in land, professional training and recruitment. Nevertheless, this huge expenditure does not guarantee a great return, and it is possible to suffer a substantial loss if the entrepreneur makes a wrong decision. Provision of home fittings is one of the risks that the housing entrepreneur faces. Some homebuyers may dislike the fittings and thus decide not to purchase the home. In some places, such as Hong Kong, land supply is scarce, but demand is huge (seven million people live in a small city with a hilly landscape). Land price occupies a relatively large proportion of the costs of dwelling production. Provision of fittings amounts to only a small proportion of the total costs of construction. Entrepreneurs thus focus on land purchases more than provision of fittings. The risk of installing fittings is not very high. Furthermore, many developers have run their business for many years and have accumulated sufficient knowledge on buyers' taste. Provision of fittings to customers is common. Yet, in mainland China, especially in areas of low land costs, costs of fittings have become relatively high. The risk of supplying fittings that prospective owners dislike is higher and this leads to the phenomenon of the bare flats' sale in China's real estate market.

A housing shortage reduces the pressure on developers to provide more than the most basic dwellings. Hence, it increases the probability that housing developments are sold in absence of fittings. In the other hand, changes in per capita income does not change this tendency in any significant way.

All in all, real estate developers have a higher tendency to build bare units when:

1   There is a shortage of housing so that developers do not need to face intense competition and supply extra stuff, i.e. fittings in addition to the building structure itself;
2   Land costs are low; the costs of providing fittings become relatively high and developers tend not to take the risk of speculating in consumers' taste of fittings;
3   There were no repeated sales during the observation period, hence consumers have little guide to predict the quality of fittings.
4   Finally, there is an insignificant relationship between the proportion of bare units construction and income. This chapter has also demonstrated that the Probit model is a powerful and appropriate tool for this type of analysis.

Acknowledgement

This chapter is an extended and revised version of "Econometric modeling of risk adverse behaviors of entrepreneurs in the provision of house fittings in China," *Construction Economics and Building,* Vol. 12, No. 1, 72–82 (2012) (previously known as the *Australasian Journal of Construction Economics and Building*).

3.8 References

Baltas, G. and C. Saridakis (2009) Brand-name Effects, Segment Differences, and Product Characteristics: An Integrated Model of the Car Market. *Journal of Product and Brand Management*, 18, 143–151.
Basham, P. and J. Luik (2011) The Social Benefit of Gambling. *Economic Affairs*, 31, 9–13.
Bechern, S. I. (2008) Asymmetric Information in Consumer Contracts: The Challenge That Is Yet to Be Met. *American Business Law Journal*, 45, 723–774.
Berkelaar, A. B., R. Kouwenberg and T. Post (2004) Optimal Portfolio Choice under Loss Aversion. *The Review of Economics and Statistics*, 86, 973–987.
Blanchflower, D. G. and A. J. Oswald (1998) What Makes an Entrepreneur? *Journal of Labor Economics*, 16, 26–60.

Brue, S. L., C. R. McConnell and S. M. Flynn (2009) *Essentials of Economics*. New York: McGraw Hill.

Burnett, J. and H. R. Bruce (2007) New Consumers Need New Brands. *Journal of Product and Brand Management*, 16, 342–347.

Carland, J. W., F. Hoy, W. R. Boulton and J. A. C. Carland (1984) Differentiating Entrepreneurs from Small Business Owners: A Conceptualization. *The Academy of Management Review*, 9, 354–359.

Chau, K. W., S. K. Wong and C. Y. Yiu (2007) Housing Quality in the Forward Contracts Market. *The Journal of Real Estate Finance and Economics*, 34(3), 313–325.

Chen, H. L. (2007) Gray Marketing and Its Impacts on Brand Equity. *Journal of Product and Brand Management*, 16, 247–256.

Chongqing, S. I. O. (2008) *Chongqing Statistical Yearbook*.

Das, T. K. and B. S. Teng (2001) Trust, Control, and Risk in Strategic Alliances: An Integrated Framework. *Organization Studies*, 22, 251–283.

Drydakis, N. (2011) Ethnic Discrimination in the Greek Housing Market. *Journal of Population Economics*, 24, 1235–1255

Du, H., B. Li, J. Zuo and R. Y. M. Li (2014) The Optimal Principal–Agent Model for the CO2 Allowance Allocation under Asymmetric Information. *Hong Kong Shue Yan University Working Paper*.

Fairoz, F. M., T. Hirobumi and Y. Tanaka (2010) Entrepreneurial Orientation and Business Performance of Small and Medium Scale Enterprises of Hambantota District Sri Lanka. *Asian Social Science*, 6, 34–46.

Francis-Brophy, E. and J. Donoghue (2013) Social Housing Policy Challenges in Tasmania. *Australian Journal of Social Issues*, 48, 435–454.

Genesove, D. and C. Mayer (2001) Loss Aversion and Seller Behavior: Evidence from the Housing Market. *The Quarterly Journal of Economics*, 116, 1233–1260

Granlund, D. and N. Rudholm (2011) Consumer Information and Pharmaceutical Prices: Theory and Evidence. *Oxford Bulletin of Economics and Statistics*, 73, 230–254.

Grewal, D., R. Krishnan, J. Baker and N. Borin (1998) The Effect of Store Name, Brand Name and Price Discounts on Consumers' Evaluations and Purchase Intentions. *Journal of Retailing*, 74, 331–352.

Gujarati, D. N. (2006) *Essentials of Econometrics*. Boston: McGraw-Hill/Irwin.

Gwin, C. R. and S. E. Ong (2000) Homeowner Warranties and Building Codes. *Journal of Property Investment and Finance,* 18, 456–472.

Hangzhou Soufun (2009) *Hangzhou Soufun*.

Hangzhou Statistics Bureau (2008) *Hangzhou Statistical Yearbook*.

Hassan M. R., B. Nath and M. Kirley (2007) A Fusion Model of HMM, ANN and GA for Stock Market Forecasting. *Expert Systems with Applications*, 33, 171–180

Hawthorne, L. and B. Birrell (2002) Doctor Shortages and their Impact on the Quality of Medical Care in Australia. *People and Place,* 10, 55–67.

Horppu, M., O. Kuivalainen, A. Tarkiainen and H. K. Ellonen (2008) Online Satisfaction, Trust and Loyalty, and the Impact of the Offline Parent Brand. *Journal of Product and Brand Management,* 17, 403–413.

Ingersoll, R. M. and T. M. Smith (2003) The Wrong Solution to Teacher Shortage. *Education Leadership,* 60, 30–33.

Jaafar, M., A. R. Nuruddin and S. P. S. A. Bakar (2014) Business Success and Psychological Traits of Housing Developers. *Construction Economics and Building,*14, 57–72.

Janiszewski, C. (1993) Preattentive Mere Exposure Effects. *Journal of Consumer Research*, 20, 376–392.

Jensen, M. and R. Drozdenko (2008) The Changing Price of Brand Loyalty Under Perceived Time Pressure. *Journal of Product and Brand Management*, 17, 115–120.

Jin, X. H. and H. Doloi (2008) Interpreting Risk Allocation Mechanism in Public–Private Partnership Projects: An Empirical Study in Transaction Cost Economics Perspective. *Construction Management and Economics*, 26, 707–721.

Laforet, S. (2011) Brand Names on Packaging and their Impact on Purchase Preference. *Journal of Consumer Behaviour*, 10, 18–30.

Lai, L. W. C. and P. Y. L. Chan (2004) The Formation of Owners' Corporations in Hong Kong's Private Housing Estates: A Probit Evaluation of Mancur Olson's Group Theory. *Property Management*, 22, 55–68

LeBel, J. L. and N. Cooke (2008) Branded Food Spokescharacters: Consumers' Contributions to the Narrative of Commerce. *Journal of Product and Brand Management*, 17, 143–153.

Li, H., F. F. Tang, L. Huang and F. Song (2009) A Longitudinal Study on Australian Online DVD Pricing. *Journal of Product and Brand Management*, 18, 60–67.

Li, R. Y. M. (2009a) Legal Regulations, People's Perceptions on Law and Scope of Services Provided by Firms: A Study on Dwelling Developers in Beijing and Hong Kong. *The Business Review, Cambridge*, 12, 221–226.

Li, R. Y. M. (2009b) The Myth of Fly-by-night Developers in Shanghai and Harbin. *Economic Affairs*, 29, 66–71

Li, R. Y. M. (2010a) A Study on the Impact of Culture, Economic, History and Legal Systems which Affect the Provisions of Fittings by Residential Developers in Boston, Hong Kong and Nanjing. *International Journal of Global Business and Management Research*, 1, 131–141

Li, R. Y. M. (2010b) *Factors Influencing Developers' Decision to Sell Housing Units With Fittings – Empirical Evidence from China*. The University of Hong Kong: Unpublished Ph.D. Thesis.

Li, R. Y. M. (2011a) *Building Our Sustainable Cities*. Illinois: Common Ground Publishing.

Li, R. Y. M. (2011b) *Everyday Life Application of Neo-institutional Economics* Lambert Academic Publishing.

Li, R. Y. M. (2011c) Imperfect Information and Entrepreneurs' Choice on Provision of House Fittings. *Lex Et Sciential Economic Series*, 18, 269–276.

Li, R. Y. M. (2014a) *Law, Economics and Finance of the Real Estate Market – A Perspective of Hong Kong and Singapore*. Germany: Springer.

Li, R. Y. M. (2014b) Transaction Costs, Firms' Growth and Oligopoly: Case Studies in Hong Kong Real Estate Agencies' Branch locations. *Asian Social Science*, 10, 40–52.

Li, R. Y. M. (2015) *Construction Safety and Waste Management: An Economic Analysis*. Germany: Springer.

Li, R. Y. M. and K. W. Chau (2014) Housing Fittings and Facilities in China. In *International Conference on Banking, Real Estate and Financial Crises: Hong Kong, China and the World*.

Li, R. Y. M. and S. W. Poon (2011) Using Web 2.0 to Share the Knowledge of Construction Safety as a Public Good in Nature Among Researchers: the Fable of Economic Animals. *Economic Affairs*, 31, 73–79.

Matzler, K., S. G. Kräuter and S. Bidmon (2008) Risk Aversion and Brand Loyalty: the Mediating Role of Brand Trust and Brand Affect. *Journal of Product and Brand Management*, 17, 154–162.

Noruzi, M. R., J. H. Westover and G. R. Rahimi (2010) An Exploration of Social Entrepreneurship in the Entrepreneurship Era. *Asian Social Science*, 6, 3–10.

Olson, E. L. (2008) The Implications of Platform Sharing on Brand Value. *Journal of Product and Brand Management,* 17, 244–253.

Omar, M., R. L. Williams and D. Lingelbach (2009) Global Brand Market-Entry Strategy to Manage Corporate Reputation. *Journal of Product and Brand Management,* 18, 177–187.

Pan, M. C., Kuo, C. Y. and M. C. Pan (2015) Measuring the Effect of Chinese Brand Name Syllable Processing on Consumer Purchases, *Internet Research*, 25, 150–168.

Pechtl, H. (2008) Price Knowledge Structures Relating to Grocery Products. *Journal of Product and Brand Management,* 17, 485–496.

Pires, C., M. Vigario and A. Cavaco (2015) Brand Names of Portuguese Medication: Understanding the Importance of Their Linguistic Structure and Regulatory Issues/ Nomes de Marca dos Medicamentos Portugueses: sua Estrutura Linguistica e Aspetos Regulatorios, *Ciência and Saúde Coletiva*, 20, 2569–2583.

Pitta, D. A. and F. J. Franzak (2008) Foundations for Building Share of Heart in Global Brands. *Journal of Product and Brand Management,* 17, 64–72.

Polić-Vižintin, M., D. Štimac, Z. Šostar and I. Tripković (2014) Distribution and Trends in Outpatient Utilization of Generic Versus Brand Name Psychopharmaceuticals During a Ten-Year Period in Croatia. *BMC Health Services Research*, 14, 343–352.

Powell, M. and D. Ansic (1997) Gender Differences in Risk Behaviour in Financial Decision-Making: An Experimental Analysis. *Journal of Economic Psychology,* 18, 605–628.

Preece, C. and F. Kerrigan (2015) Multi-Stakeholder Brand Narratives: An Analysis of the Construction of Artistic Brands. *Journal of Marketing Management*, 31, 1207–1230.

Richard, S. F. (1975) Multivariate Risk Aversion, Utility Independence and Separable Utility Functions. *Management Science,* 22, 12–21.

Ricketts, M. (2015) Adverse Selection, Gresham's Law and State Regulation. *Economic Affairs*, 35, 109–122.

Shanghai Soufun (2009) *Shanghai Soufun.*

Shanghai Statistics Bureau (2008) *Shanghai Statistics Yearbook 2008.*

Shannon, R. and R. Mandhachitara (2008) Causal Path Modeling of Grocery Shopping in Hypermarkets. *Journal of Product and Brand Management,* 17, 327–340.

Soufun, 2009. 104 Cities. http://china.soufun.com/

Souiden, N. and F. Pons (2009) Product Recall Crisis Management: The Impact on Manufacturer's Image, Consumer Loyalty and Purchase Intention. *Journal of Product and Brand Management,* 18, 106–114.

Tovar, P. (2009) The Effects of Loss Aversion on Trade Policy: Theory and Evidence. *Journal of International Economics*, 78, 154–167.

Valentinov, V. (2008) The Transaction Cost Theory of the Nonprofit Firm: Beyond Opportunism. *Nonprofit and Voluntary Sector Quarterly*, 37, 5–18.

Vukasovič, T. (2009) Searching for Competitive Advantage with the Aid of the Brand Potential Index. *Journal of Product and Brand Management,* 18, 165–176.

Wang, C. X., S. Webster and N. C. Suresh (2009) Would a Risk Averse Newsvendor Order Less at a Higher Selling Price? *European Journal of Operational Research*, 196, 544–553.

Wang, Y. P. and A. Murie (1996) The Process of Commercialisation of Urban Housing in China. *Urban Studies,* 33, 971–989.

Williams, D. R., W. J. Duncan and P. M. Ginter (2006) Structuring Deals and Governance after the IPO: Entrepreneurs and Venture Capitalists in High Tech Start-ups. *Business Horizons,* 49, 303–311.

Wooldridge, J. M. (2003) *Introductory Econometrics: A Modern Approach*. Australia: South-western College Publications.

Wu, W. (1999) Reforming China's Institutional Environment for Urban Infrastructure Provision. *Urban Studies,* 36, 2263–2282.

Xiao, T. and D. Yang (2008) Price and Service Competition of Supply Chains with Risk averse Retailers under Demand Uncertainty. *International Journal of Production Economics* 114, 187–200.

Zahirovic-Herbert, V. and S. Chatterjee (2012) In Search of Value: How Subdivision Names Influence House Prices and Marketing Duration. *Housing and Society* 39, 51–76.

Zilg, A. (2011) THAT'S AMORE: Brand Names in the Italian Food Market. *International Journal of Applied Linguistics* 21, 1–25.

Zou, P. X. W., G. Zhang and J. Wang (2007) Understanding the Key Risks in Construction Projects in China. *International Journal of Project Management,* 25, 601–614.

4 Forecasting real estate stock prices in Hong Kong

A State Space model approach

*Rita Yi Man Li, Yikun Huang and
Kwong Wing Chau*

4.1 Introduction

Direct real estate investment in, for example, housing, shops, car parks . . . offers a golden chance for global investors to reap a huge profit. The direct real estate market suffers from various problems – it is obsolete, indivisible, the requirement of huge sums of money, complicated taxation policies and so on – so many small investors invest in the indirect real estate markets instead. Moreover, owing to higher liquidity, smaller transaction costs, the existence of a public market place, larger number of market participants, the indirect real estate market is more efficient than the direct real estate market (Li and Li 2011). Hence, there is more indirect investment in recent years and the expansion of indirect investment has also led to individuals focusing on the indirect real estate market: such as, property stocks and REITs (Li and Chow 2015, Li and Li 2011).

Econometric forecasting has been applied in various areas such as GDP, unemployment rate, housing supply and demand. In the financial area, forecasting stock price is important in both academic and industrial research. Accurate stock price forecasting helps investors make appropriate decisions on investment. Many different methods have been used in the last few decades (Hassan *et al.* 2007). Nevertheless, the prediction of stock price performance needs to consider interactions of a number of variables and can be complex (Yoon and Swales, 1991). Some of them suggested that stock price fluctuations were the result of rumors about future price fluctuations (Kenneth 1988). Others documented the impact of changes in real economic variables, such as real gross national product, inflation, industrial production, interest rates and money supply on stock returns (Liow *et al.* 2006).

In line with the efficient market hypothesis (EMH), Pearce and Roley (1985) pointed out that the anticipated components of economic announcements did not affect stock price movements significantly. Piotroski and Roulstone (2004) also argued that a large portion of variation in stock return was not caused by market and industry movements. The so-called New Economy – with higher labor productivity growth and the information technology revolution – has been identified as an important factor which not only powers economic growth (Li 2011) but can also drive stock prices to a historically high level (Balke and Wohar 2002).

In spite of the fact that many of the investors consider that the decision making process of stock purchase is rational. Baker *et al.* (2003), however, argued that stock price changes did not reflect rational human behavior. It did reflect a certain level of irrationality. Apart from industry and market, casual intuition and psychological evidence suggested that sunny weather was associated with upbeat mood and there was a significant correlation between sunshine and stock returns (Hirshleifer and Shumway 2003).

In Hong Kong, stock prices dropped when a famous pop star Adam Cheung's program was on TV, on a number of occasions. Nevertheless, Adam is an actor only. We may consider the first couple of downswings in stock prices when Adam's TV program was on to be coincidence. However, this is likely to become a self-fulfilling prophecy. Another example of a self-fulfilling prophecy as an example of investors' psychological impact on the stock market can be found in May, June and July nearly every year in Hong Kong. There is a widely held superstitious believe that stocks are poor in May and June and recover in July (in Chinese 五窮六絕七翻身). Such a believe turns out to be true for many years in Hong Kong except some years such as 2015. Nevertheless, the phenomenon itself is not supported by any changes in macroeconomic factors or business sectors changes.

Partly because there is no consensus on the line of reasoning which leads to stock price fluctuations, prediction and forecasting attract many researchers' attention (Hsu *et al.* 2009). Forecasting using the disclosure of corporate information such as sales volume, profit to earnings ratio and so on was a topic which often leads to intensive debate within the investment community (Patell 1976). On top of that, the mythological success of Warren Buffet raises questions on the precise prediction of the fundamental stock price. Penman (1992) was of the view that accounting earnings are one of the major considerations in performing stock price analysis. This was consistent with the viewpoint made by Abarbanell and Bushee (1997), that financial statement data provides the fundamental signals for stock performance. This chapter aims to forecast stock prices with the help of individual companies' financial statement data.

The major problem of forecasting by using financial statement data rests on the infrequent release data. For example, GDP is released once for each quarter and population data is released every five years. In the same vein, corporates only release financial statements semi-annually, stock price data is available every day. Even though some stock prices are relatively stable, semi-annual information is insufficient to forecast stock prices. It means that the independent variables do not match with the dependent variable. Hence, a higher frequency of company and macro economic data is needed.

This chapter applies the State Space model, which was originally developed in engineering to simulate the data which lies between that of the Internet – such as, the data from real estate companies – and macro economic data, which can be obtained from the Hong Kong Census and Statistics. Property stocks in Hong Kong are chosen in our research because the real estate industry and related stocks contribute a significant proportion of the stock market in Hong Kong: the property and construction companies' stocks reach 45% of total stock market capitalization. Accordingly, changes in property prices often have great impact on the Heng Seng Index and the overall stock market in Hong Kong.

To invest in the real estate market, modern investors can put their money in physical form: a hotel, a shopping mall, an industrial building or office. Alternatively, they can invest in indirect ones, such as real estate property companies' stock, derivatives or Real Estate Investment Trusts (REITs) (Liow and Huang 2006) (Clayton 2007; Ong and Ng 2009). Although previous studies indicate that direct real estate property performs better than indirect property in terms of return and risk (Chiang and Ganesan 1996), the liquid nature of stocks attracts a large crowd of investors (Tse 2001). Sometimes, speculators also invest in the stock market to hedge against possible loss due to other investment tools' poor performance. Partly because of its high liquidity in nature and relatively low capital requirement, many investors choose to invest in stocks.

4.2 Hong Kong stock market and real estate stocks

Compared to other stock markets nearby, such as Shenzhen and Shanghai, the Hong Kong stock market has a relatively long history. The first formal stock exchange – The Association of Stockbrokers in Hong Kong – was formed in 1891. The fast growth of the Hong Kong economy led to the birth of three other stock exchange markets: the Far East Exchange, the Kam Ngan Stock Exchange and the Kowloon Stock Exchange (Hong Kong Exchanges and Clearing Limited 2009). The Stock Exchange Ordinance in 1973 was implemented to restrict the number of stock markets to four (Hsu *et al.* 2006). Nevertheless, the co-existence of many stock markets in Hong Kong did not last long. In 1980, they were unified into the Stock Exchange of Hong Kong (Hong Kong Exchanges and Clearing Limited 2009). Since 1986, the stock market in Hong Kong has developed rapidly in terms of market liquidity and capitalization (Zhu and Liow 2005), ranking third after the stock markets in London and New York.

Stock markets do not only provide a golden chance for individual investors to reap a profit with a small sum of money, it also provides a good channel for large companies to obtain funds cheaply. As developers perceived the stock market as a good source of finance for land purchase and corporate development, many of them became listed companies gradually. For example, Cheung Kong, Sun Hung Kai, Sino became listed companies in 1972, followed by Henderson Land (1981), China Overseas (1992), China Resources Land (1992). As more and more developers listed, the contribution of property company stocks to Hong Kong stock market also increased. Prior to 1995, construction and property and company stocks contributed around one-fourth to Hong Kong stock market capitalization; which was, however, greater than other Western countries and South-East Asia substantially. The contribution of property and construction companies' stocks amounted to 45% of total stock market capitalization (Newell and Chau, 1996). The significance of property companies on the Hong Kong stock market is reflected in the fact that six of the top ten listed companies is property or heavily property-related companies (Zhu and Liow 2005).

Although the previous research results suggested that there may not be a direct linkage between the product of a company and the corresponding stock prices. For example, Roll 1984's research found that there was no linkage between the price of orange juice and the share prices (Roll 1984). Linkage between real estate markets has caught the attention of academics in the 1990s. As price fluctuations in the physical property and stock markets share similar factors – GDP, unemployment rate, business activities etc. – the performance of different real estate markets displays similar ups and downs. By applying econometric analysis, Zhu and Liow (2005) concluded that there is a long-term and short term linkage between different property stock markets. For example, research based on the Error Correction Model and the Johansen cointegration test shows a long-term contemporaneous relationship between the Hong Kong and Shanghai property stock markets (Zhu and Liow 2005).

4.3 A literature review of the stock price forecast method

As many investors are interested in making a profit, in order to satisfy the needs of investors, there is also an intense interest in stock price forecasting. Indeed, there is an extensive literature with regards to this. By using the data from the US from 1872 to 2007, empirical results show that the power of out-of-sample bootstrap tests varies from 0.16 to 0.27 (Wu and Hu 2011, forthcoming). By using the historical data of ACI pharmaceutical company data from 31 August 2010 to 30 September 2010 to predict future stock values with Feed forward Neural Networks, the average error of the first simulation was 3.71% (Khan *et al.* 2011). The Takagi–Sugeno–Kang (TSK) type Fuzzy Rule Based System has also been used to develop stock price prediction. The TSK fuzzy model applies 1) a technical index as the input variables and 2) a consequent linear combination of the input variables. A fuzzy rule-based model on the Taiwan Electronic Shares from the Taiwan Stock Exchange (TSE) successfully forecasted the price variation for stocks from different sectors to an accuracy level of 98.08% in MediaTek and 97.6% in the TSE index (Chang and Liu 2008).

By using Neural Networks, Kolarik and Rudorfer (1994)'s research results recorded a forecast error of 7.97%, much lower than the traditional forecast method Autoregressive Integrated Moving Average (ARIMA). Tsang *et al.* 2007's research adopted the neural network method to forecast the stock prices of Hong Kong and Shanghai Banking Corporation (HSBC) Holdings and achieved an overall hit rate of more than 70% (Tsang *et al.* 2007). Although the State Space model has been widely adopted in forecasting agricultural product prices (Vukina and Anderson 1993), inflation (Burmeister *et al.* 1986) and traffic flow (Stathopoulos and Karlaftis 2003), there is little overseas research which adopts State Space for forecasting stock prices: Aoki and Havenner (1991) performed their forecast according using US data. In view of this, this chapter aims at testing the forecast capability of the State Space model when applied to Hong Kong Property stocks. In the following sections we will present a detailed explanation of the model, the data used, the results and our discussions.

4.4 Research method

The traditional Dividend Discount Model considers that today's stock price equals the present value of all expected future dividends. Such that P0 = present value of $(DIV_1, DIV_2, DIV_3, \ldots\ldots DIV_t)$ (Brealey *et al.* 2012):

$$P_0 = \frac{DIV_1}{(1+r)} + \frac{DIV_2}{(1+r)^2} + \frac{DIV_3}{(1+r)^3} + \ldots + \frac{DIV_t + P_t}{(1+r)^t}$$

In view of this, our proposed State Space model uses companies' financial statement data, which has a substantial connection with the future stock price, as well as the dividend.

By assuming a constant growth rate of dividend, as well as infinite dividend payment, we have the Gordon Dividend Growth Model as the following:

$$P_0 = \frac{DIV_1}{k_e - g}$$

Hence, the stock price can be determined by the first dividend payment, cost of equity and growth rate of dividend.

We can obtain the abovementioned three components from the financial statements of the various companies. After that, they can be determined by other basic information. For example, the profitability of a company will turn out to be the asset available for dividend payment. Also, the asset volume and structure will decide the required rate of return for equity, as the marginal cost of equity increases. Furthermore, the growth of the dividend should be reasonably judged through financial information. Additionally, the financial statement terms are a reflection of corporate operation, financial activities and investment activities. Such activities are taken in certain economic conditions so that we may use macroeconomic variables to represent the items in a financial statement.

In summary, we have a three-layered logical structure for stock price evaluation. The stock price is our core target, while the financial statement data are in the middle and the economic environment comprises the outer layer.

4.4.1 The State Space model

A State Space model is a relational structure which characterizes a system's behavior. It represents all the possible transitions inside the system and the corresponding states (Pelánek 2008). This model has been applied in the control theory literature since the 1960s (Allen and Pasupathy 1997). Ziemer *et al.* (1998) concur that Classical control theory is usually based on the Laplace Transform. Suppose the simple dynamic system follows the equation:

$$\frac{dY^M(T)}{dT^M} + A_1 \frac{dY^{M-1}(T)}{dT^{M-1}} + \ldots + A_M Y(T) = B_0 \frac{dX^N(T)}{dT^N} + \ldots + B_N X(T)$$

To determine Y(T) for given u(T), Fourier transform f input signal:

$$X(f) = \int_{-\infty}^{\infty} X(T) e^{-2j\pi ft} dT$$

where $\left| e^{-2j\pi ft} \right| = 1$.

Use single-side Fourier transform of X(T):

$$\int_{0}^{\infty} \left(X(T) e^{-\sigma T} \right) e^{-JWT} dT = \int_{0}^{\infty} X(T) e^{-(\sigma+JW)T} dT$$

Followed by Laplace transform:

$$L[X(T)] = \int_{-\infty}^{\infty} X(T) e^{-ST} dT (S = \sigma + JW)$$

Yet, modern control theory is based on the state equations. Ziemer *et al.* (1998) provides a detailed illustration on the State Space model. State equations consist of two parts: state equation and output equation. The detailed equations can be shown according to the following equations:

$$\begin{cases} AX + BV = \dot{X} \text{ transition equation} \\ CX + DV_v = y \text{ output equation} \end{cases}$$

where:

X: state vector, $x = (x_1, \ldots x_n)^T$
$V = (v_1, \ldots, v_m)^T$: input vector
$Y = (y_1, \ldots y_p)^T$: output vector

The transition equation shows how the state space changes from time to time. The change of state space, which is the term $\dot{X}$, depends on previous state and the input variables; and the output equation provide the information about how state space and input variables determines the output of the whole system. Matrix A, B, C and D are parameters to be estimated and we assume both functions are linear. Figure 4.1 shows how the system works under the State Space model:

4.4.2 Model development for stock price forecasting according to the State Space model

To apply the State Space model for stock forecasting, we adopt the stock price as the output variable. According to Penman (1992), Abarbanell and Bushee (1997), the stock price reflects data available in financial statement. Therefore, variables from financial statement are used in the State Space model for stock price forecast. They

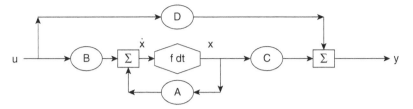

Figure 4.1 Diagram of State Space model

are represented by the X vector in the abovementioned model. Further, we assume that input variables do not affect the stock price directly so that matrix D above is 0. The output function for stock price forecasting is Y = CX.

All the variables in state space are financial variables available from individual companies. The macro economy provides an environment which affects property companies' performance; for example, an increase in inflation may raise the nominal value and profit in the financial statement. In view of the above, the transition equation for stock price forecast is: $AX + BV = \dot{X}$, where $\dot{X}$ is the data from financial statement while V is the data from macro-environment. The state change $\dot{X}$ refers to the difference in state space variables between two periods. Assume that the function is linear, the State Space model for stock price forecasting is:

$$\begin{cases} X_t - X_{t-1} = AX_{t-1} + BV_{t-1} \\ \qquad Y_t = CX_t \end{cases}$$

The monthly stock price forecast is based on less frequent data than that available to the market, such as firms' biannual financial statement and macroeconomic data: the State Space model allows us to generate the missing data by transition equation. While Allen and Pasupathy (1997) adopt the State Space model to handle the low frequency data problem in the real GDP growth rate forecast, this chapter adopts the same idea – using the transition equation in the State Space model to forecast the stock price.

4.5 Data description

The Hong Kong Hang Seng Index is comprised of five sub-indices (commerce, finance, property, utilities and industry). At present, there are seven real estate companies' shares which have been included in the Hang Seng Index – Property. Details are illustrated in Table 4.1. Monthly stock transaction data from January 2003 to October 2010 of these seven companies was obtained from Yahoo Finance.

Although these seven properties are included in the same Heng Seng Sub-index Properties, they have a lot of differences with regards to performances and history. These seven property companies were listed in different years. The oldest

Table 4.1 Seven property companies included in Heng Seng Sub-index Properties

Stock code	Name of the corporations	Year founded	Year listed	Beta	Market cap (US$ million)	Shares outstanding (million)	Annual dividend	Yield (per cent)
1	Cheung Kong	1950	1972	0.97	261,958.20	2,316.16	2.7	2.39
12	Henderson Land	1976	1981	1.1	115,413.30	2,165.35	1	1.88
16	Sun Hung Kai Properties	1963	1972	1.09	325,115.19	2,564.00	2.7	2.13
83	Sino Land	1971	1972	1.58	75,390.41	4,889.13	0.4	2.59
101	Hang Lung Prop	1987	1954	1	150,977.80	4,159.17	0.71	1.96
688	China Overseas*	1979	1992	1.1	140,233.09	8,172.09	0.23	1.34
1109	China Resources Land*	1938	1992	1.35	85,610.79	5,030.01	0.28	1.63

Source: Heng Seng Indexes 2010; Reuters 2010.
*These have not been included in this study because the majority of their businesses rest in China; using Hong Kong data for forecasting is therefore not suitable.

was China Resources Land which was listed in 1938, followed by Cheung Kong Holdings in 1950 and Sun Hung Kai in 1963. The latest listed properties company Hang Lung Properties had been listed in Hong Kong stock market since 1987. Cheung Kong recorded the highest annual dividend at 2.7 and China Overseas was the lowest at 0.4 only. Sino Land had the highest yield of 2.59% and China Overseas had the lowest yield at 1.34%.

Among the seven companies in the Hang Seng Property index, China Overseas and China Resources Land are excluded in our study because the major source of income comes from mainland China so as their businesses. Therefore, including macroeconomic data in Hong Kong for estimating their stock price is not reasonable in this sense.

In order to apply the State Space model for this empirical study, we need to specify the variables in State Space and Input Vector. We select the Gross value of the company, Semi-annual Profit and Semi-annual Turnover in the financial statement as state space variables. The reason for choosing these three variables is that: 1) Gross value is the book value for companies, and the stock price reflects the market value of the company. Therefore, they correlate strongly. 2) Turnover measures income within the period between two statements (it's a proxy for micro operation of the company). 3) Profit shows the earnings of a company. All financial statement data comes from the official website of each company.

We include the following twelve macro-variables as variables in input vector. All of them are related to the real estate industry and are collected from the Hong Kong Census and Statistics Department.

Generally speaking, two types of macro variables are included in the model. One is real estate related variables, which includes (1) Quarterly GDP for the real estate service and finances, (2) Quarterly GDP changes for the construction sector, (3) Quarterly gross value change for the construction sector. We believe that turnover, profit and gross value change of property stocks are directly related to real estate activities. Nine other macroeconomic related variables are also included. They show the macroeconomic environment for all financial markets. More importantly, most of them are useful in explaining fluctuations in the stock price index or individual stock prices. These nine variables are: 1) Quarterly GDP data, 2) Quarterly GDP growth rate, 3) Monthly money supply (M3) (Rogalski and Vinso, 1977), 4) Monthly interest rate, 5) Monthly CPI (Schwert, 1981), 6) Monthly CPI change rate (Pearce and Roley, 1985), 7) Monthly unemployment rate (Boyd, Hu and Jagannathan, 2005), 8) Monthly Heng Seng Index (HSI), 9) Monthly HSI change rate. The following table shows the summary of the twelve variables selected.

4.6 Empirical study with the State Space model and results of the research

We applied the State Space model to do the forecast. The State Space model is applied individually to each of the property company; and regression method is used to fit the following State Space model:

$$\begin{cases} X_t - X_{t-1} = AX_{t-1} + BV_{t-1} \\ \quad\quad Y_t = CX_t \end{cases}$$

Table 4.2 Definition of input variables for the State Space model design in this chapter

Variable	Definition	Frequency	Related research which has adopted this variable
GDP_re	GDP change rate for real estate sector	Quarterly	
GDP_cont	GDP change rate for construction sector	Quarterly	
GV_cont	change rate for gross value of construction	Quarterly	
GDP	Gross Domestic Product	Quarterly	
GDP_rate	GDP growth rate	Quarterly	
M3	Money supply	Monthly	Rogalski and Vinso, 1977
INT	Interest rate	Monthly	
CPI	Consumer Price Index	Monthly	Schwert, 1981
CPI_rate	CPI change rate	Monthly	Pearce and Roley, 1985
Unemp	Unemployment rate	Monthly	Boyd, Hu and Jagannathan, 2005
HIS	Heng Seng Index	Monthly	
HSI_rate	HSI change rate	Monthly	

Because many variables can be included in the State Space model, variables selection has been applied to reduce the number of variables. As we have a relatively small sample size, we use forward variable selection method. The following criteria have been applied in variable selection: 1) the new added variable should be significant at 10% level and increase the adjusted R-square after selecting it. 2) If more than two variables are selected under the first criterion, we choose the one which provides the largest increase in adjusted R-square.

We then obtain the output equations relation so that we can get the significant variables in state space. After that, we work on the transition equations according to the significant variables. The following table shows the result of output equation estimation. Only significant variables are recorded here.

After determining the Output equations, we need to find out the State Space transition equations to do forecast. As we have mentioned, the transfer function describes how state space move from time to time. Such dynamic relationship provides a recursive method for working on forecasting.

One problem in working on the transition equations is that new variables may be added to the state space above: we need to enlarge the state space by adding the new variables. One example in our study is the turnover of Sun Hung Kai Properties. Only Profit and Gross value are chosen. Nevertheless, when we select the variables to explain the dynamic change of profit and value, turnover of the company becomes significant. Thus, turnover should be included in SHK's state space. We use the ordinary least square (OLS) to fit the transition equations again for each property company, and the results are summarized in Table 4.4.

Table 4.3 Output function estimation

Company name	Variables	Coefficients	t-Statistics	Adjusted. R-square for regression
Cheung Kong	Gross Value	0.000388	18.05467	0.473194
Henderson Land	Profit	0.001187	1.933350	0.704809
	Turnover	0.004305	2.484071	
	Gross Value	0.000138	1.868552	
Sun Hung Kai Properties	Profit	0.003224	2.790594	0.693149
	Gross Value	0.000297	4.934588	
Sino Land	Profit	0.001550	1.974302	0.581211
	Gross Value	0.000123	3.204010	
Hang Lung Property	Gross Value	0.000318	13.08972	0.664636

Table 4.4 Transfer function estimation

Company name	State Space variables	Selected input variables	Coefficients	t-Stat.	Adjusted. R-square
Cheung Kong	Gross Value	GDP	0.048265	5.090091	0.037367
Henderson Land	Profit	Const.	3459.250	2.007145	0.266993
		Profit	−0.619002	−2.237562	
	Turnover	Turnover	−1.657034	−4.437133	0.629203
		HIS	0.404574	4.329856	
	Gross Value	Gross Value	−0.773332	−2.938352	0.413582
		M3	0.016090	3.148525	
SHK Prop.	Profit	Profit	−1.851012	−8.514210	0.855824
		Turnover	0.915326	7.583578	
		GDP_RE	681.0924	4.063228	
		INT	1665.424	2.504477	
	Gross Value	Turnover	0.680503	4.878173	0.075435
	Turnover	Const.	26007.59	6.658885	0.789498
		Profit	−0.556100	−4.078309	
		Turnover	−0.824196	−4.919548	
		Unemployment	−1962.322	−4.823545	
Sino Land	Profit	Profit	−1.554309	−5.447188	0.694395
		Value	0.042843	3.615263	
		INT	1065.206	2.964756	
	Gross Value	Const.	2297.920	3.028165	0.472579
		GDP_rate	513.5986	3.534737	
		CPI_rate	−923.5774	−2.647084	
Hang Lung Prop.	Gross Value	GV_cont	−1490.355	−2.574662	0.143995

Because financial statements are released every six months, the change $X_t - X_{t-1}$ in regression of the transition equations captures six month's transition for each company. Nevertheless, as we aim at forecasting the monthly stock price, we need to obtain the monthly change for state space. Therefore, we have to decompose the State Space Variable change in these six months. For simplification, we assume that monthly changes are equal, so that the transition equation for two months is just one sixth of the total transition equation. Therefore, we can have the monthly transition relation for state space; and eventually calculate the monthly state space condition and predicted stock price.

Another point we need to mention here is the adjustment using the updated financial statement data. The recursive structure of the State Space model provides a tool for forecasting continuously if the initial value is given. However, to precisely obtain the stock price prediction, we consider the information which can be obtained from the updated financial statements. In spite of the fact that the transition and output equations are estimated by regression of all the data, the fluctuation structure can be eliminated by regression. In the light of these, we adopt the adjustment every six months with the new financial statement.

As our aim is to use the property stocks data to forecast the stock price, accurate prediction is of the utmost importance. To be exact, the accuracy of our forecast is calculated by the following equation:

$$\text{Error rate} = \frac{\text{forecast} - \text{actual}}{\text{actual}}$$

We expected the average error rate to be close to zero in the case of a good forecast. The following table shows the average error rate of forecasting value for the five selected property companies in Hong Kong. All of them, except Sino Land, achieve a forecast error of less than 10% (a lot better than that in the Neutral Network method (Tsang *et al.* 2007)). For a detailed result, please refer to the Appendix.

4.7 The State Space model forecasts better for large scale companies than small companies. Why?

4.7.1 The first reason: the nature of the State Space model

The result shows that the average error rate is positively related to the variance of error rate of the stock price, i.e., the more volatile the stock price, the more difficult for us to predict it precisely. One major reason for this is that the State Space model mainly uses infrequent data to construct the state space, while the missing value for state space is calculated by frequent data from input variables. Therefore, the estimation of state space maintains an infrequent change in structure. When stock performance is volatile, the predictability of infrequent change structure reduces so that the average error rate increases. When we perceive the issue from another angle, the forecasting values from our State Space model are more likely to be the long-term fundamental prices for stocks, while

Table 4.5 Forecasting results

Company name	Market cap (US$ million)	Average error rate	Variance of error rate
Cheung Kong	261,958.20	4.28%	0.040822206
Henderson Land	115,413.30	3.45%	0.039815806
Sun Hung Kai Properties	325,115.19	0.61%	0.027681231
Sino Land	75,390.41	20.94%	0.153157357
Hang Lung Properties	150,977.80	8.80%	0.070075293

short-term fluctuations are likely to be captured. But the short-term deviation will always move along the fundamental prices.

Among the five companies selected, four of them have the average error rate below 10%, and three of them can reach 5% error level. This result supports the use of the State Space model for forecasting. Therefore, the State Space model recovers the missing financial statement data because of the statement's low frequency, and such recovery is quite accurate according to the frequent output stock price.

4.7.2 The second reason: relationship between firm size and volatility

Cheung Kong Holdings and Sun Hung Kai are relatively large in scale, like Henderson Land. Hang Lung is the fourth base on current value and Sino is the fifth. The State Space model did a great forecast job for Sun Hung Kai and Henderson with average error rate of less than 5%. Nevertheless, the error rate for Henderson Land and Sino was substantially higher. Henderson Land's average error rate was around 8%, with Sino over 20%. Sino's market capitalization value was US$75,390.41, which is one-tenth of the other four firms' capitalization value only.

As mentioned above, previous research shows that stock volatility is negative related to company size, which means the stock price for small companies tends to be more volatile. As the State Space model we construct here is based on the fundamental variables available from financial statement, volatility of stock price increases the error of forecasting. Therefore, the State Space model performs better for large scale real estate firms.

4.7.3 The third reason: hot money, speculation and small companies

In many places, many overseas investors earn short term interest rates or make profit by investing the money in another countries temporarily, leading to the inflow of hot money in some countries. Hence, apart from the aforementioned two reasons, we also speculate that the hot money effect is more significant for small companies which leads to a larger fluctuation in stock price. Because of speculation and the insufficient volume of trading, deviation of stock price cannot be adjusted immediately.

4.8 Conclusion

As investors wish to make profit, many of them would like to know if there is any crystal ball in the market to help them do the forecast and earn more money. Partly because of this reason, those who work in ivory towers are also looking for various means to do the forecast so as to satisfy practical needs. Moreover, accurate forecasting has academic values. In a world of imperfect information, stock investors always need to estimate future stock prices using limited and non-frequent data. The differences in data frequency further increases the difficulty of making accurate forecasts.

This chapter has attempted to forecast the monthly property stock price based on limited non-frequent data, released in different time frames: some bi-annually, others quarterly. The State Space model can forecast monthly stock price movements closely; the average error rate can be kept below 10% for large scale property stock (measured in terms of market capitalization rate). We consider that the high error rate for Sino is mainly due to the fact that small scale firms' stocks are more volatile than large ones'. Estimation is therefore more difficult.

4.9 References

Abarbanell, J. S. and B. J. Bushee, (1997) Fundamental Analysis, Future Earnings, and Stock Prices. *Journal of Accounting Research*, 35, 1–24.

Allen, D. S. and M. Pasupathy (1997) *A State Space Forecasting Model with Fiscal and Monetary Control*. URL http://research.stlouisfed.org/wp/1997/97-017.pdf.

Aoki, M. and A. Havenner (1991) State Space Modeling of Multiple Time Series. *Econometric Reviews*, 10, 1–59.

Baker, M., J. C. Stein and J. Wurgler (2003) When Does the Market Matter? Stock Prices and the Investment of Equity-Dependent Firms. *The Quarterly Journal of Economics*, 118, 969–1005.

Balke, N. S. and M. E. Wohar (2002) Low-Frequency Movements in Stock Prices: A State Space Decomposition. *The Review of Economics and Statistics*, 84, 649–667.

Boyd, J. H., J. Hu and R. Jagannathan (2005) The Stock Market's Reaction to Unemployment News: Why Bad News Is Usually Good for Stocks. *The Journal of Finance*, 60, 649–672.

Brealey, R. A., S. C. Myers and A. J. Marcus (2012) *Fundamental of Corporate Finance*. McGraw Hill, New York.

Burmeister, E., K. D. Wall and J. D. Hamilton (1986) Estimation of Unobserved Expected Monthly Inflation Using Kalman Filtering. *Journal of Business* and *Economic Statistics*, 4, 147–160.

Chang, P. C. and C. H. Liu (2008) A TSK Type Fuzzy Rule Based System for Stock Price Prediction. *Expert Systems with Applications*, 34, 135–144.

Chiang, Y. H. and S. Ganesan (1996) Property Investment in Hong Kong. *Journal of Real Estate Portfolio Management*, 2, 141–158.

Clayton, J. (2007) Commercial Real Estate Derivatives: The Developing US Market. *Journal of Economic Perspectives*, 32, 33–40.

Heng Seng Indexes (2010) *HSI–Properties*. URL http://www.hsi.com.hk/HSI-Net/HSI-Net?cmd=navigation&pageId=en.indexes.hsisis.hsi.constituents.

Hirshleifer, D. and T. Shumway (2003) Good Day Sunshine: Stock Returns and the Weather. *The Journal of Finance*, 58, 1009–1032.

Hong Kong Exchanges and Clearing Limited (2009) *A Glimpse of the Past*. URL http:// www.hkex.com.hk/eng/exchange/corpinfo/history/documents/hkex-01e.pdf.

Hsu, B., D. Arner, M. Tse and S. Johnstone (2006) *Financial Markets in Hong Kong Law and Practice*. Oxford University Press, London.

Hsu, S. H., J. P. A. Hsieh, T. C. Chih and K. C. Hsu (2009) A Two-stage Architecture for Stock Price Forecasting by Integrating Self-organizing Map and Support Vector Regression. *Expert Systems with Applications* 36, 7947–7951.

Kenneth, D. W. (1988) Dividend Innovations and Stock Price Volatility. *Econometrica*, 56, 37–61.

Khan, Z. H., Alin T. S. and M. A. Hussain (2011) Price Prediction of Share Market using Artificial Neural Network (ANN). *International Journal of Computer Applications*, 22, 42–47.

Kolarik, T. and G. Rudorfer (1994) Time Series Forecasting Using Neural Networks. *Time Series and Neural Networks*, 86–94.

Li, R. Y. M. (2011) Internet Boost the Economic Growth of mainland China? Discovering Knowledge from Our World Wide Web. *Global Business and Management Research: An International Journal*, 3, 345–355.

Li, R. Y. M. and Ho Ping Chow (2015) An Economic Analysis on REITs Cycles in Nine Places. *Real Estate Finance*, 32, 23–28.

Li, R. Y. M. and Y. L. Li, (2011) An Overview of Regulations That Protect Real Estate Stocks, REITs and Derivative Investors in Hong Kong. *Lex Et Sciential Economic Series*, 18, 200–211.

Liow, K. H. and Q. Huang (2006) Interest Rate Risk and Time-Varying Excess Returns for Asian Property Stocks. *Journal of Property Investment & Finance*, 24, 188–210.

Liow, K. H., M. F. Ibrahim and Q. Huang (2006) Macroeconomic Risk Influences on the Property Stock Market. *Journal of Property Investment & Finance*, 24, 295–323.

Newell, G. and K. W. Chau, (1996) Linkages between Direct and Indirect Property Performance in Hong Kong. *Journal of Property Finance*, 7(4), 9–29.

Ong, S. E. and K. H. Ng (2009) Developing the Real Estate Derivative Market for Singapore: Issues and Challenges. *Journal of Property Investment and Finance*, 27, 425–432.

Patell, J. A. (1976) Corporate Forecasts of Earnings Per Share and Stock Price Behavior: Empirical Test. *Journal of Accounting Research*, 14, 246–276.

Pearce, D. K. and V. V. Roley (1985) Stock Prices and Economic News. *The Journal of Business*, 58, 49–67.

Pelánek, R. (2008) Properties of State Spaces and Their Applications. *International Journal of Software Tools Technology Transfer*, 10, 443–454.

Penman, S. H. (1992) Return to Fundamentals. *Journal of Accounting, Auditing and Finance*, 7, 452–482.

Piotroski, J. D. and B. T. Roulstone (2004) The Influence of Analysts, Institutional Investors, and Insiders on the Incorporation of Market, Industry, and Firm-Specific Information into Stock Prices. *The Accounting Review*, 79, 1119–1151.

Reuters (2010) Developers' Coporate Stock Information. URL http://www.reuters.com

Rogalski, R. J. and J. D. Vinso (1977) Stock Returns, Money Supply and the Direction of Causality. *The Journal of Finance*, 32, 1017–1030.

Roll, R. (1984) Orange Juice and Weather. *American Economic Review* 74, 861–880.

Schwert, G. W. (1981) The Adjustment of Stock Prices to Information about Inflation. *The Journal of Finance*, 36, 15–29.

Stathopoulos, A. and M. G. Karlaftis (2003) A Multivariate State Space Approach for Urban Traffic Flow Modeling and Prediction. *Transportation Research Part C*, 11, 121–135.

Tsang, P. M., P. Kwok, S. O. Choy, R. Kwan, S. C. Ng, J. Mak, J. Tsang, K. Koong and T. L. Wong (2007) Design and Implementation of NN5 for Hong Kong Stock Price Forecasting. *Engineering Applications of Artificial Intelligence*, 20, 453–461.

Tse, R. Y. C. (2001) Impact of Property Prices on Stock Prices in Hong Kong. *Review of Pacific Basin Financial Markets and Policies*, 4, 29–43.

Vukina, T. and J. L. Anderson (1993) A State-space Forecasting Approach to Optimal Intertemporal Cross-hedging. *American Journal of Agricultural Economics*, 75, 416–424.

Wu, J.L. and Y. H. Hu (2012) Price–Dividend Ratios and Stock Price Predictability. *Journal of Forecasting,* 31, 423–442.

Yoon, Y. and G. Swales (1911) *Predicting Stock Price Performance: A Neural Network Approach*. Proceedings of the 24th Annual Hawaii International Conference on System Sciences, 156–162.

Zhu, H. and K. H. Liow (2005) Relationship Between the Shanghai and Hong Kong Property Stock Markets *Pacific Rim Property Research Journal*, 11, 24–44.

Ziemer, R. E., W. H. Tranter, D. R. Fann and R. E. Ziemer (1998) *Signals and Systems: Continuous and Discrete*. Prentice Hall, Upper Saddle River, N.J.

Appendix I: Macroeconomic data set used for this chapter

Date	Real estate GDP	Construction GDP	GDP	GDP change rate	Money supply M3	INT	CPI	CPI change rate (annual)	Unemployment	Stock	Stock return	Gross value of construction
Mar-03	-0.7	-2.9	299,502	-1.1	3,548,561	0.19	101.20	-1.90	7.80	9005.35	-0.02	-3.3
Jun-03	-5.2	-7	287,670	-7.5	3626092	0.14	100.30	-2.47	8.70	9260.57	0.05	-1.1
Sep-03	-1.5	-3	317,747	-3.4	3712282	0.11	98.40	-3.67	8.07	10757.90	0.06	0.7
Dec-03	-0.4	-6.9	329,842	-1.4	3858044	0.05	99.30	-2.33	7.07	12361.17	0.03	-5.1
Mar-04	-1.9	-9.2	308,143	2.9	3858504	0.04	99.33	-1.87	7.03	13292.69	-0.04	-5.1
Jun-04	-3.1	-10.7	310,687	8.0	3875750	0.15	99.40	-0.83	6.93	12142.32	0.01	-10.1
Sep-04	-5.8	-12	327,651	3.1	3948801	0.43	99.20	0.80	6.77	12736.11	0.02	-6.2
Dec-04	0.3	-2.3	345,442	4.7	4189544	0.44	99.73	0.20	6.03	13781.62	0.02	0.5
Mar-05	3.1	0.5	321,331	4.3	4189596	0.61	99.70	0.37	5.83	13811.31	0.00	1.7
Jun-05	6.2	-8.4	331,485	6.7	4192238	1.32	100.13	0.83	5.90	13992.37	0.02	-2.4
Sep-05	5.1	-5.5	356,460	8.8	4270883	2.12	100.43	1.43	5.47	15071.02	-0.01	-2.9
Dec-05	-0.1	-12.4	373,314	8.1	4407188	2.86	101.03	1.27	4.90	14733.31	0.03	-5.5
Mar-06	-1.2	-12.6	349,049	8.6	4551810	3.04	101.30	1.57	4.93	15825.55	0.02	-6.4
Jun-06	-1.9	-4.6	350,588	5.8	4666865	3.09	102.17	2.07	5.13	16262.27	0.01	-0.6
Sep-06	0.8	-10.9	375,411	5.3	4886502	3.06	102.73	2.30	4.63	17302.22	0.03	-3.0
Dec-06	4.2	-3.1	400,310	7.2	5089741	2.90	103.20	2.17	4.13	19083.18	0.03	1.9
Mar-07	6.7	6.1	373,686	7.1	5302796	2.82	103.03	1.73	4.27	19852.95	0.00	-1.0
Jun-07	9.1	3.8	378,215	7.9	5624126	2.86	103.50	1.27	4.43	20908.73	0.04	0.0
Sep-07	7.8	-2.8	415,498	10.7	5975806	2.88	104.40	1.57	3.93	24770.52	0.10	0.1
Dec-07	12.2	-2.6	448,056	11.9	6139758	2.62	106.77	3.47	3.13	29269.61	-0.10	3.1
Mar-08	12.8	9.2	410,438	9.8	6017338	1.36	107.73	4.57	3.27	23545.54	0.03	4.3
Jun-08	2.9	-0.3	402,082	6.3	5944671	0.80	109.40	5.73	3.40	24130.16	-0.04	2.9
Sep-08	-1.8	-1.5	428,891	3.2	6091633	0.93	109.23	4.63	3.67	20669.73	-0.16	-3.5
Dec-08	-6.2	2.8	433,905	-3.2	6300751	0.79	109.23	2.33	4.30	14081.46	-0.02	-2.7
Mar-09	-4.9	-3.4	381,670	-7.0	6265738	0.48	109.57	1.70	5.30	13221.93	0.05	-6.4
Jun-09	1.6	-1	390,465	-2.9	6514058	0.37	109.33	-0.10	5.73	17356.91	0.09	-0.8
Sep-09	5.2	7	414,516	-3.4	6665330	0.21	108.27	-0.87	5.27	20417.59	0.02	0.2
Dec-09	6.2	4.7	445,633	2.7	6626843	0.17	110.70	1.33	4.50	21815.62	-0.03	-6.5
Mar-10	0.5	3.5	420,314	10.1	6623605	0.15	111.67	1.93	4.53	20656.68	0.02	-0.5

Appendix II: Detail result of forecasting

Company	Cheung Kong			Henderson Land		
Date	Actual	Forecasted	Error rate	Actual	Forecasted	Error rate
Jan-03	42.69			17.95		
Feb-03	41.02			17.67		
Mar-03	36.08			15.55		
Apr-03	36.08			15.74		
May-03	40.85			17.86		
Jun-03	40.25	65.85718	0.63620323	18.14	24.04495473	0.325521209
Jul-03	44.08	66.15646579	0.500827264	20.71	24.26668688	0.171737657
Aug-03	52.03	66.48704307	0.277859755	24	24.74462888	0.031026203
Sep-03	52.68	66.81762034	0.268367888	26.28	25.44007921	−0.031960456
Oct-03	56.01	67.14819761	0.198860875	26.52	26.13465388	−0.014530397
Nov-03	52.98	67.49135826	0.273902572	25.77	27.0813741	0.050887625
Dec-03	53.2	67.839084	0.275170752	28.33	26.1551102	−0.076769848
Jan-04	62.49	68.18224465	0.091090489	32.63	26.81377703	−0.178247716
Feb-04	64.22	68.50283013	0.066689974	32.71	27.64304601	−0.15490535
Mar-04	56.44	68.8234156	0.219408497	30.81	28.5513433	−0.073309208
Apr-04	51.9	69.14400107	0.332254356	29.18	28.90472463	−0.0094337
May-04	51.8	69.46723327	0.341066279	28.26	29.0109948	0.02657448
Jun-04	50.92	69.682472	0.368469599	28.01	26.86482281	−0.040884584
Jul-04	50.92	70.0057042	0.374817443	28.6	27.54542359	−0.036873301
Aug-04	59.77	70.34658538	0.17695475	31.76	28.13599128	−0.114106068
Sep-04	59.33	70.68746656	0.19142873	31.1	28.84404002	−0.072538906
Oct-04	57.44	71.02834775	0.236565943	30.18	29.51672281	−0.021977375
Nov-04	66.56	71.38773831	0.072532126	34.22	30.13586822	−0.119349263
Dec-04	69.23	69.90402	0.009735953	34.22	23.38204566	−0.316714037
Jan-05	63.67	70.26341056	0.103556001	31.18	25.6633948	−0.176927684
Feb-05	66.34	70.59771655	0.064180231	30.93	27.26344407	−0.118543677
Mar-05	61.44	70.93202254	0.154492554	29.32	28.67774164	−0.021905128
Apr-05	65.23	71.26632853	0.092539147	30.94	29.57105767	−0.044245065
May-05	64.68	71.61119853	0.107161387	30.34	30.42710704	0.00287103
Jun-05	68.54	73.684692	0.075061161	31.8	38.03820174	0.196169866
Jul-05	76.12	74.029562	−0.027462402	33.68	38.25173795	0.135740438
Aug-05	76.53	74.40041545	−0.027826794	33.12	38.5635555	0.16435856
Sep-05	79.53	74.77126891	−0.059835673	33.12	38.71675485	0.168984144
Oct-05	73.62	75.14212237	0.020675392	29.53	38.92659546	0.318205061
Nov-05	73.62	75.53051038	0.02595097	29.78	38.72434299	0.300347313
Dec-05	72.66	73.690124	0.014177319	31.73	33.78902688	0.064892117
Jan-06	75.99	74.078512	−0.025154468	34.3	35.12699395	0.024110611
Feb-06	74.8	74.44165522	−0.004790706	36.21	36.38674539	0.004881121
Mar-06	74.94	74.80479843	−0.001804131	37.39	37.31698958	−0.001952672
Apr-06	79.73	75.16794164	−0.057218843	39.96	37.92098356	−0.051026437
May-06	78.1	75.53268599	−0.032872138	35.53	38.65017202	0.087817957
Jun-06	78.19	77.505716	−0.008751554	35.26	38.6080166	0.094952258
Jul-06	78.19	77.87046035	−0.004086707	37.37	38.99426999	0.043464543
Aug-06	79.81	78.26103003	−0.019408219	38.55	39.56383581	0.026299243

Sep-06	77.68	78.65159971	0.01250772	38.42	40.10654446	0.043897565
Oct-06	79.12	79.04216939	−0.000983703	37.72	40.54198713	0.074814081
Nov-06	85.74	79.45864346	−0.073260515	37.98	41.18065627	0.08427215
Dec-06	89.43	79.804228	−0.107634709	38.7	39.5230025	0.021266214
Jan-07	96.1	80.22070207	−0.165237231	40.26	40.77911436	0.012894048
Feb-07	89.33	80.6094771	−0.097621436	39.45	41.83510748	0.060458998
Mar-07	92.27	80.99825212	−0.122160484	40.61	42.48226364	0.046103513
Apr-07	95.54	81.38702715	−0.148136622	42.25	43.02676911	0.018385068
May-07	95.94	81.78051405	−0.147586887	48.45	43.73279998	−0.09736223
Jun-07	97.17	85.938896	−0.115582011	47.82	38.9105068	−0.186313116
Jul-07	105.23	86.3323829	−0.179583931	51.14	41.54424019	−0.187637071
Aug-07	108.64	86.76465825	−0.201356239	48.94	44.04990896	−0.099920127
Sep-07	121.34	87.1969336	−0.281383438	55.36	46.1501561	−0.166362787
Oct-07	143.71	87.62920894	−0.390235829	61.69	48.72139822	−0.210222107
Nov-07	138.76	88.09535695	−0.365124265	62.25	52.03336611	−0.164122633
Dec-07	137.33	89.476292	−0.348457788	67.82	64.341861	−0.051284857
Jan-08	119.63	89.94244	−0.248161498	61.64	63.0368461	0.022661358
Feb-08	113.35	90.36945103	−0.202739735	58.24	60.55961983	0.039828637
Mar-08	105.16	90.79646206	−0.136587466	51.98	58.98379652	0.134740218
Apr-08	115.54	91.22347309	−0.210459814	56.38	57.2987493	0.01629566
May-08	115.97	91.64179072	−0.209780196	51.41	56.94358896	0.107636432
Jun-08	101.66	91.720872	−0.097768326	46.05	46.656423	0.013168795
Jul-08	106.59	92.13918962	−0.135573791	46.24	47.70150274	0.031606893
Aug-08	108.52	92.58539877	−0.146835618	45.01	48.69808772	0.081939296
Sep-08	83.87	93.03160791	0.109235816	32.22	48.93158962	0.51867131
Oct-08	70.55	93.47781705	0.324986776	25.68	48.03475229	0.870512161
Nov-08	70.99	93.92924265	0.323133436	25.6	46.136603	0.802211055
Dec-08	71.28	91.376328	0.281935017	27.94	39.183335	0.402409986
Jan-09	70.89	91.8277536	0.295355531	29.35	38.81814435	0.322594356
Feb-09	62.09	92.22483501	0.485341198	25.51	38.29290957	0.501094064
Mar-09	64.91	92.62191642	0.426928307	28.82	37.86000651	0.313671288
Apr-09	79.15	93.01899783	0.17522423	35.84	37.88435985	0.057041291
May-09	95.33	93.42522938	−0.01998081	45.71	38.70668619	−0.153211853
Jun-09	88.46	92.526748	0.045972733	43.35	38.479833	−0.11234526
Jul-09	99.5	92.93297954	−0.066000206	50.42	40.42921531	−0.198151223
Aug-09	91.25	93.36423324	0.023169679	44.58	42.65768756	−0.043120512
Sep-09	98.4	93.79548693	−0.046793832	48.9	44.0434405	−0.099316145
Oct-09	100.8	94.22674063	−0.065210906	54.84	45.49114466	−0.170475116
Nov-09	97.7	94.69036779	−0.030804833	53.82	46.82313083	−0.130005001
Dec-09	100.7	95.992752	−0.046745263	57.86	48.42273195	−0.16310522
Jan-10	92.05	96.45637917	0.04786941	48.98	49.12974302	0.003057228
Feb-10	94.9	96.89366498	0.021008061	51.79	49.08387593	−0.052251865
Mar-10	100	97.33095079	−0.026690492	54.16	49.22150079	−0.091183516
Apr-10	97	97.7682366	0.007919965	49.6	49.53755611	−0.001258949
May-10	88.75	98.19839686	0.10646081	46.35	49.73355137	0.07300003
Jun-10	90.8	98.439092	0.084130969	46	49.44654123	0.074924809
Avg. Error rate			0.042794009			0.034519382

(continued)

Appendix II: (continued)

Company	SHK Properties			Hang Lung Properties		
Date	Actual	Forecasted	Error rate	Actual	Forecasted	Error rate
Jan-03	36.76			5.69		
Feb-03	35.94			5.46		
Mar-03	31.08			5.69		
Apr-03	30.41			5.27		
May-03	32.32			5.74		
Jun-03	32.66	45.316969	0.387537324	5.47	7.8988656	0.444033931
Jul-03	38.47	45.60035609	0.185348482	6.41	7.985763674	0.245828966
Aug-03	46.95	47.19828285	0.00528824	7.03	7.9304649	0.12808889
Sep-03	52.14	48.42687709	−0.071214479	7.66	7.875166125	0.028089572
Oct-03	54.22	49.41251899	−0.088666193	7.93	7.819867351	−0.013888102
Nov-03	52.73	50.67124526	−0.039043329	7.72	8.222758421	0.065124148
Dec-03	54	45.588898	−0.155761148	8	7.7056488	−0.0367939
Jan-04	63.28	47.08398373	−0.255942103	9.53	8.10853987	−0.149156362
Feb-04	64.54	47.9515712	−0.257025547	9.05	8.51143094	−0.059510393
Mar-04	60.63	48.88133623	−0.19377641	8.94	8.91432201	−0.002872258
Apr-04	57.01	49.81036485	−0.126287233	8.53	9.31721308	0.092287583
May-04	56.8	50.3121446	−0.114222806	8.24	10.11509539	0.227560121
Jun-04	54.46	53.758367	−0.012883456	8.16	10.1676366	0.246033897
Jul-04	56.16	55.50019037	−0.011748747	8.73	10.96551891	0.256073186
Aug-04	61.69	55.79675463	−0.095529995	9.22	11.45530806	0.242441221
Sep-04	62.55	56.28576651	−0.100147618	9.34	11.9450972	0.27891833
Oct-04	61.27	56.75652135	−0.073665393	9.38	12.43488635	0.325680847
Nov-04	66.62	59.42970207	−0.10793002	10.11	12.39538722	0.226052149
Dec-04	67.05	59.33056	−0.115129605	10.03	15.9028302	0.585526441
Jan-05	62.53	60.0746979	−0.039265986	9.73	15.86333108	0.630352629
Feb-05	62.74	61.92990927	−0.01291187	10.11	15.72903405	0.555789718
Mar-05	61.61	63.64475648	0.0330264	9.59	15.59473703	0.626145676
Apr-05	64.88	65.22385585	0.005299874	10.1	15.46044001	0.530736634
May-05	64.66	67.9716973	0.051217094	9.55	15.6500358	0.638747205
Jun-05	66.84	60.424984	−0.095975703	9.67	12.9245694	0.336563537
Jul-05	69.88	65.30180683	−0.065515071	10.43	13.1141652	0.25735045
Aug-05	68.49	68.94285399	0.006611972	10.22	13.32746047	0.304056797
Sep-05	69.88	71.88156999	0.028642959	10.39	13.55655539	0.304769528
Oct-05	63.92	74.2687241	0.161901191	9.73	13.78565031	0.416819148
Nov-05	65.59	74.5251918	0.136227958	10.17	14.22014068	0.398243922
Dec-05	67.28	90.738378	0.348667925	10.56	13.1903856	0.249089545
Jan-06	71.23	87.48946387	0.228267077	13.18	13.62487597	0.033753867
Feb-06	72.08	84.72908356	0.175486731	12.22	14.13046476	0.156339179
Mar-06	70.57	82.77120905	0.172895126	13.04	14.63605356	0.122396745
Apr-06	79.3	81.46673147	0.027323222	13.74	15.14164235	0.102011816
May-06	72.18	80.43375859	0.114349662	12.07	15.1890413	0.258412701
Jun-06	70.97	71.440811	0.006633944	12.38	14.0839338	0.13763601
Jul-06	73.21	72.95615141	−0.003467403	13.52	14.13133275	0.045216919
Aug-06	75.99	75.34748523	−0.008455254	14.66	14.3683275	−0.019895805

Sep-06	76.12	77.36838012	0.016400159	14.64	14.60532224	−0.002368699
Oct-06	76.17	79.14129888	0.039008781	14.98	14.84231699	−0.009191122
Nov-06	80.34	81.96776065	0.020260899	15.62	14.69222032	−0.059396907
Dec-06	81.43	88.574431	0.087737087	17.53	20.0291982	0.142566925
Jan-07	86.13	89.44392964	0.038475904	19.23	19.87910153	0.03375463
Feb-07	84.4	91.09311231	0.079302279	18.6	19.95809978	0.073016117
Mar-07	82.92	92.36508102	0.113905946	19.8	20.03709802	0.011974648
Apr-07	84.02	93.30253796	0.1104801	21.2	20.11609627	−0.051127534
May-07	83.34	94.90211259	0.138734252	22.34	20.11609627	−0.099548063
Jun-07	86.5	89.557751	0.035349723	24.42	17.756484	−0.272871253
Jul-07	92.1	93.64538599	0.016779435	26.55	17.756484	−0.331205876
Aug-07	95.69	95.96542056	0.002878259	25.87	17.74858418	−0.313931806
Sep-07	120.3	97.56831573	−0.188958306	31.53	17.74068435	−0.437339539
Oct-07	134.62	98.70152765	−0.26681379	33.74	17.73278453	−0.474428437
Nov-07	149.77	101.1590865	−0.324570431	32.41	17.48788995	−0.460416848
Dec-07	153.57	107.220671	−0.301812392	33.19	19.599453	−0.409477162
Jan-08	142.06	106.4327179	−0.250790385	28.42	19.35455843	−0.318981055
Feb-08	129.52	105.9405112	−0.182052878	26.36	19.01486596	−0.278646967
Mar-08	112.67	105.2124276	−0.066189513	25.4	18.67517349	−0.264756949
Apr-08	127.63	104.7884636	−0.178966829	29.27	18.33548102	−0.373574273
May-08	117.25	100.9778755	−0.138781445	27.38	18.10638609	−0.338700289
Jun-08	98.92	112.664249	0.138943075	23.05	21.8308908	−0.05288977
Jul-08	109.39	106.8635175	−0.023096101	22.91	21.60179588	−0.057101882
Aug-08	100.61	101.3898709	0.007751426	23.05	21.87828975	−0.050833417
Sep-08	73.3	97.8776931	0.335302771	16.49	22.15478362	0.343528418
Oct-08	61.24	95.79584145	0.564269129	17.55	22.43127749	0.27813547
Nov-08	59.31	93.13708178	0.570343648	15.93	22.64457276	0.421504882
Dec-08	62.45	67.545363	0.081591081	16.07	21.7992498	0.356518345
Jan-09	67.67	73.0101042	0.078913909	16.98	22.01254507	0.296380746
Feb-09	58.97	77.6184405	0.316236061	14.29	22.51813387	0.575796632
Mar-09	68.02	81.10051315	0.192303927	17.6	23.02372266	0.30816606
Apr-09	78.88	83.69093675	0.060990577	21.35	23.52931145	0.102075478
May-09	94.38	87.97942683	−0.06781705	24.86	23.59251005	−0.050985115
Jun-09	94.97	98.8776	0.041145625	24.24	26.7595092	0.103940149
Jul-09	114.92	100.0729623	−0.12919455	27.55	26.8227078	−0.026398991
Aug-09	102.2	101.7704308	−0.004203221	23.18	26.80690815	0.156467133
Sep-09	111.89	102.5932057	−0.083088697	27.55	26.7911085	−0.027545971
Oct-09	116.68	102.9912111	−0.117319069	29.21	26.77530885	−0.083351289
Nov-09	113.65	103.5502367	−0.088867253	27.55	27.28879747	−0.009481036
Dec-09	115.44	117.648772	0.019133507	29.94	27.819594	−0.070821844
Jan-10	100.45	112.4536408	0.119498664	26.37	28.33308262	0.074443785
Feb-10	106.9	106.7803833	−0.001118959	29.34	28.37258174	−0.032972674
Mar-10	116.7	102.9462068	−0.117855983	30.77	28.41208087	−0.076630456
Apr-10	109.6	100.4643546	−0.083354429	28.31	28.45157999	0.005001059
May-10	102.7	98.94363319	−0.036576113	26.93	28.49107912	0.057968033
Jun-10	107.5	98.1245352	−0.087213626	29.83	28.53057824	−0.043560904
Avg. Error rate			0.006085353			0.087961846

(continued)

Appendix II: (continued)

Company	Sino Land		
Date	Actual	Forecasted	Error rate
Jan-03	1.95		
Feb-03	1.95		
Mar-03	1.8		
Apr-03	1.74		
May-03	1.86		
Jun-03	2.03	4.159937426	1.049230259
Jul-03	2.45	4.645876487	0.896276117
Aug-03	3.05	5.075506554	0.66410051
Sep-03	3.51	5.42593808	0.545851305
Oct-03	3.41	5.697785428	0.670904818
Nov-03	3.41	5.922136458	0.736696908
Dec-03	3.72	5.227312662	0.405191576
Jan-04	4.54	5.446727593	0.199719734
Feb-04	4.46	5.668104447	0.270875437
Mar-04	4.1	5.881363528	0.434478909
Apr-04	4.01	6.081586825	0.516605193
May-04	3.86	6.316007292	0.636271319
Jun-04	3.69	6.039614443	0.636751882
Jul-04	4.16	6.343785109	0.524948343
Aug-04	4.92	6.550641178	0.331431134
Sep-04	4.92	6.799237223	0.381958785
Oct-04	5.64	7.013378504	0.243506827
Nov-04	6.39	7.216063657	0.129274438
Dec-04	6.6	7.860935654	0.191050857
Jan-05	5.96	7.915443392	0.328094529
Feb-05	5.96	8.000016794	0.342284697
Mar-05	6.13	8.125900119	0.325595452
Apr-05	6.37	8.298414429	0.302733819
May-05	6.9	8.574953901	0.242746942
Jun-05	7.2	12.07332965	0.67685134
Jul-05	7.98	11.54872414	0.447208538
Aug-05	7.5	11.33169597	0.510892797
Sep-05	8.15	11.26147003	0.381775464
Oct-05	7.46	11.28095212	0.512191973
Nov-05	8.09	11.4254655	0.412294871
Dec-05	8.31	7.948052486	−0.043555658
Jan-06	9.54	9.022825156	−0.0542112
Feb-06	10.29	9.853450317	−0.042424653
Mar-06	9.85	10.51764434	0.067781152
Apr-06	11.43	11.05057601	−0.03319545
May-06	10.54	11.45326666	0.086647691
Jun-06	10.98	15.02909464	0.368770003
Jul-06	11.6	14.60685118	0.259211309
Aug-06	11.62	14.30179548	0.230791349
Sep-06	12.24	14.08719441	0.150914576
Oct-06	11.99	13.95515746	0.163899705
Nov-06	13.21	13.90126429	0.052328863

Dec-06	16.41	9.861202538	−0.399073581
Jan-07	15.83	10.75557687	−0.320557367
Feb-07	16.25	11.45334313	−0.295178884
Mar-07	15.24	12.03064439	−0.210587639
Apr-07	15.04	12.45984508	−0.171552854
May-07	15.53	12.83378407	−0.173613389
Jun-07	14.82	15.40881172	0.039730885
Jul-07	16.05	15.06794289	−0.061187359
Aug-07	16.54	14.87520156	−0.100652868
Sep-07	17.56	14.77634768	−0.158522342
Oct-07	21.84	14.74278849	−0.324963897
Nov-07	25.31	14.72619158	−0.418167065
Dec-07	25.35	16.4889221	−0.349549424
Jan-08	21.59	16.15835828	−0.251581367
Feb-08	18.37	15.78363878	−0.140792663
Mar-08	15.43	15.26396778	−0.010760351
Apr-08	18.2	14.86621064	−0.18317524
May-08	18.36	14.48158759	−0.211242506
Jun-08	14.32	13.96109704	−0.025063056
Jul-08	14.5	13.83381029	−0.045944118
Aug-08	12.77	13.71886502	0.07430423
Sep-08	7.85	13.65795572	0.73986697
Oct-08	6.01	13.64092161	1.269704095
Nov-08	5.49	13.60026803	1.477280152
Dec-08	7.72	12.53901909	0.624225271
Jan-09	7.31	12.72597901	0.740900001
Feb-09	5.86	12.75234031	1.176167288
Mar-09	7.47	12.78014764	0.710863138
Apr-09	9.73	12.77860921	0.313320576
May-09	14.07	12.80928146	−0.089603308
Jun-09	12.5	12.00283178	−0.039773458
Jul-09	15.4	12.25155043	−0.204444777
Aug-09	12.94	12.43715545	−0.038859703
Sep-09	13.51	12.5739218	−0.069287801
Oct-09	15.06	12.65001942	−0.160025271
Nov-09	14.64	12.73908588	−0.12984386
Dec-09	15.08	14.78796636	−0.019365626
Jan-10	12.89	14.40384233	0.11744316
Feb-10	14.66	14.22059249	−0.029973227
Mar-10	15.28	14.09768388	−0.077376709
Apr-10	14.2	14.05681664	−0.010083335
May-10	12.82	14.01508367	0.093220255
Jun-10	14.12	14.00453218	−0.008177607
Avg. Error rate			0.20944468

5 Market sentiment and property prices in Hong Kong

A heteroscedasticity-and-auto correlation-consistent approach

Rita Yi Man Li and Joe Fuk Kin Wong

5.1 Introduction

Basic economic studies suggest that demand for a product is determined by population, economy and cost of production. Similar to other commodities, housing prices are affected by a bundle of factors such as business cycles (Li 2014), weather (Li 2011), environmental externalities (Li 2012b) and climate (Li 2009). Chau *et al.* (2001)'s research showed that housing prices in Hong Kong are determined by the capital markets and macroeconomics factors. The impact of bad news is only reflected on the more liquid property stock prices and later in housing prices and therefore housing price appeared to lag behind indirect real estate (Newell and Chau 1996). Lee (2009)'s research conceded that a lower unemployment rate increased housing prices. The negative coefficient of the previous volatility of unemployment rates determines current volatility of housing prices.

Apart from local issues, various international financial issues may also affect local housing prices. The World housing market is flat and getting flatter. The close relationship of trading and financial markets between different places implies no one single country can stand alone as an island: financial and economic shocks which originate from one place may affect others. Lai *et al.* (2006) have shown that the negative impact on housing prices of the 1997 Asian Financial Crisis was stronger than that of SARS in 2003. Others showed that the financial crisis affected housing prices due to higher unemployment rate, lower affordability, lower expected future prices and decrease in demand for housing (Li 2012a, Li and Wong 2013, Li and Chan 2013, Li and Li 2012).

Hossain and Latif (2007)'s impulse responses analysis showed that the growth rate of gross domestic product and inflation were the major determinants of Canadian housing price volatility. Karantonis and Ge (2007) proposed that the real interest rate, real household income, dwelling completions and speculative investment affect housing prices in Sydney. Hui and Nga suggested that (2009) a reduction in real interest rate and household disposable income might lead to a rapid increase in housing. When the interest rate was low, the costs of borrowing dropped, homebuyers borrowed more which stimulated the housing demand and housing prices increased in turn. Yanga and Turner (2004) proposed that the influx of hot money inflated the housing bubble in China.

5.2 News, market sentiment and commodity prices

There is a wealth of literature on the impact of news on products' prices. Some shows that negative news about a company affects consumer behavior and perceptions (Sago and Hinnenkamp 2013). Nguyen and Claus (2013) suggested that consumers only react to bad news but not good news. Reduction in consumer sentiment decreased consumption while rises had no impact on property purchase. Their research implied that there was an asymmetric reaction to these two types of news.

On the other hand, Soroka (2006) suggested that good news increases consumption: bad news decreased the intention to purchase more than positive news. Beck and Bhagat (1997) found that negative news, such as the firm being sued, led to poor price performance.

Furthermore, Sago and Hinnenkamp (2014) pointed out that negative corporate news adversely affected consumer's affinity to favorite brands and other consumer behavior variables such as price levels, brand perception, willingness to pay and purchase. In Germany, negative news reduced the confidence of investors: they would buy less stock and the stock price decreased (Lucke 2013). Steeley (2004) suggested that official macroeconomic statistics, such as inflation rates, the money supply, unemployment and labor market variables, interest rates, government debt and unemployment significantly affected the stock market index. The significance of the news could affect distributions; a big surprise usually made a greater impact.

Good news led to an increase in the share prices (Milgrom 1981). The US economic expansion in the US in turn positively affected the Vietnam stock market (Nguyen 2011). Krishnamurti *et al.* (2013) suggested that both good news and bad news affected stock prices. Nevertheless, good news had more significant impact on stock prices. Furthermore, the stock prices might be lower when there is good news. In sharp contrast, however, bad news had a smaller impact on stock prices. When there was a negative market signal, investors might not sell shares, stock prices might be even higher. Hassan (2011)'s research showed that both the negative and positive market sentiment affected the price volatility of oil prices but the negative news affected the price in a larger extent than the positive ones.

5.3 The impact of market sentiment on real estate markets

Positive and negative market sentiment may affect real estate prices. In Hong Kong, positive market sentiment favored the marketing of their presale unit. A negative sentiment revealed a cautious view about future real estate price trends (Chau *et al.* 2010). McCollough and Karani (2014) revealed that homeowners create a downward, self-fulfilling prophecy with respect to housing prices as they fed off media reports to acquire new information about housing values. When a sudden downward trend in housing prices was reported by the media, homeowners react to these reports by readjusting their own perceptions of housing values downward. These perceptions led to a real decline in home values and sales; further, homeowners decreased their perceptions on future property prices.

In Turkey, property prices were mainly affected by the news: it affected investors' investment plans and choices (Xu and Wang 2012). The news was written
by some property sellers, who disclosed information at the same time. Research
shows that the property price declined by 4% after the seller disclosed some bad
information from the US (Pope 2008). However, Schwann and Chau (2003) found
no asymmetry effect of good and bad news on price discovery behavior between
direct and indirect real estate. Berry and Dalton (2004) suggested that the policy
interventions were negative news to the housing market, leading to the decrease in
property prices. Lee (2009) suggested that housing prices are asymmetric in good
and bad news: bad news had a stronger impact than good. He also found out that
the lending rate has little impact on the housing price.

5.4 Robert Shiller and Eugene Fama's theory: can the trend of the asset price be predicted?

The 2013 Nobel Prize in Economic Sciences was awarded to Eugene Fama,
Lars Peter Hansen and Robert Shiller for their empirical analysis of asset prices.
Fama and Shiller's theories are considered to be contradictory with regards to the
prediction of asset prices. Fama is known as the father of the "efficient markets
hypothesis" (EMH) and the origin of this idea dated back to his influential paper
in 1970. Fama (1970) classified his work on market efficiency into three categories in an attempt to answer three different questions:

1   How well did past returns predict future returns? (weak-form tests);
2   How quickly did security prices reflect public information announcements?
     (semi-strong-form tests) and;
3   Did any investors have private information that is not fully reflected in the
     market prices? (strong-form tests).

Fama (1970) further classified efficiency into weak-form, semi-strong form and
strong form. The first category, weak-form tests was revised to cover a more general
area of tests for return predictability to forecast returns with variables like dividend
yields and interest rates. With regards to the second and third categories, he changed
the title to "event studies" and "tests for private information" respectively. "Event
studies" studied the speed of adjustment of prices to firm-specific information like
dividend changes, changes in capital structure, and corporate-control transactions.
He argued that the efficiency research put forth the challenge that private information is rare. If weak-form was valid, technical analysis became ineffective.

If semi-strong form held true, one could not earn superior returns by using publicly available information. In order words, one could not "beat the market" based
on traditional security analysis and technical analysis. Under the strong-form of
the hypothesis, asset prices even swiftly reflected "private" or "insider" information. Fama's pioneer work on market efficiency gave rise to passive investment
strategies like index fund investment strategies (Fama 1991).

Following a vast amount of research conducted by Fama, many economists
now agree that historical prices are of little use in predicting the asset prices in

the short run. However, many of the controversies rest on the predictability of the asset prices in the long run.

Investigating the longer-term predictability, Shiller (1981) argued that stock prices moved much more than can be explained by dividend streams, which was contrary to the basic theory that a stock's value should equal the expected value of future dividends. Then he continued to look into the explanation of stock prices overreacted to dividends, i.e. prices were exceptionally high when dividends were high relative to recent experience and were exceptionally low when dividends are low relative to recent experience. In his paper "Stock Prices and Social Dynamics," Shiller contended that mass psychology could be an important cause of movements in the stock prices. By using the literature on social psychology, sociology, and marketing, he studied the history of the US stock market in the postwar period and found that various social movements could have major effects on the stock prices. He challenged the efficient markets hypothesis that qualitative evidence about the investors would be convincing to explain the excessive volatility of stock prices (Shiller 1984). Campbell and Shiller (2001) showed that the dividend-price and price-smoothed-earning ratios have a special significance to forecast stock prices and were extraordinarily bearish. They suggested that stock prices were substantially driven by mean reversion and observers must face the fact that something extremely unusual occurred.

Having drawn the line between "efficient markets hypothesis" and "behavioral finance," which referred to the collaboration between finance and other social sciences, Shiller (2003) claimed that:

> we have to distance ourselves from the presumption that financial markets always work well and that price changes always reflect genuine information. Evidence from behavioral finance helps us to understand, for example, that the recent worldwide stock market boom, and then crash after 2000 ("Internet" or "Dot Com" bubble) . . . . The challenge for economists is to make this reality a better part of their models.

Concerning home prices, based on the evidence of divergence between real interest rates and real rental-price ratios, he argued that there was the possibility of an irrational overpricing at that time and could pose the risk of a huge decrease in home prices in coming years (Shiller 2006). As per the discussion in his book "Irrational Exuberance" published in 2005, Shiller suggested that "there is substantial evidence that there is a strong psychological element to the current housing boom . . . the current home price boom is best thought of as a social epidemic: a fad of sorts" (Shiller 2006).

Although there are bubbles in asset prices, there were controversies on which research methods can be used to measure the turning points or ends of housing booms. Shiller (2007) researched several boom ends – the end of the stock market boom of the 1990s, the end of the California real estate boom of the 1880s, the end of the Florida land boom of the 1920s, the end of the national real estate boom of the 1980s, and the recent end of national real estate boom of the 2000s. He stressed that the change in attitudes or psychology affected home prices and the end of a housing boom might have multiple causes. Although the change in attitudes cannot be measured accurately, media and journalists' impressions could provide evidence to support the argument. He concluded that:

> ... a rising sense of enthusiasm and excitements, followed by a sense of betrayal and embarrassment at having fallen for the boom and understanding the supply response to the boom, played a significant, if unquantifiable, role in the booms and their subsequent break.

The Nobel Prize was awarded to Fama and Shiller for resolving the contradictory ideas of the predictability of asset prices over recent decades. However, we think that the controversy will continue, as reflected by the recent interview of Fama by *The New Yorker* on October 14, 2013:

Eugene Fama:	... I don't know what a credit bubble means. I don't even know what a bubble means. These words have become popular. I don't think they have any meaning.
John Cassidy:	I guess most people would define a bubble as an extended period during which asset prices depart quite significantly from economic fundamentals.
Eugene Fama:	... It's easy to say prices went down, it must have been a bubble, after the fact. I think most bubbles are twenty-twenty hindsight. Now after the fact you always find people who said before the fact that prices are too high. People are always saying that prices are too high. When they turn out to be right, we anoint them. When they turn out to be wrong, we ignore them. They are typically right and wrong about half the time.
John Cassidy:	Are you saying that bubbles can't exist?
Eugene Fama:	They have to be predictable phenomena. I don't think any of this was particularly predictable.
John Cassidy:	But what is driving that volatility (in financial prices)?
John Cassidy:	And all that is consistent with market efficiency?
Eugene Fama:	Yes. It is exactly how you would expect the market to work.
John Cassidy:	There were some people out there saying this was an unsustainable bubble (in 2007) ...
Eugene Fama:	Right. For example, (Robert) Shiller was saying that since 1996.
John Cassidy:	Yes, but he also said in 2004 and 2005 that this was a housing bubble.

Eugene Fama: . . . I didn't renew my subscription to The Economist because they use the world bubble three times on every page. Any time prices went up and down – I guess that is what they call a bubble. People have become entirely sloppy.

5.5 Mixed research method: a combination of qualitative and quantitative data analyses

This research included both qualitative content analysis and quantitative modeling. Wiser's Information portal was used to search for housing news from Oriental Daily and Apple Daily by using the keywords "housing prices" (in Chinese "樓價") in three periods:

1   From June to October 2003;
2   From January to April 2008;
3   From January to April 2013.

The first and third period was selected as they were record lows for Hong Kong housing prices; the second period was chosen as a peak was recorded. The news before and after each peak and trough was also recorded: information is not perfect and we wished to capture the period before, during and after the peak and each trough. Apple Daily and Oriental Daily were selected for this research study as these two newspapers have the largest number of readers.

Qualitative content analysis first groups similar information into one category. It creates a set of systematic and objective criteria which is then used to analyze the written texts (Li and Tsoi 2014, Li 2013). The news was then classified positive, negative and neutral.

Positive news denotes news that anticipates a housing price increase and/or a report of an increase in the housing price. Negative news refers to a gloomy expectation of housing prices and/or a report of a recent drop in housing prices. Neutral news refers to a mixture of good and bad news in the same piece of news or a report that the housing price will be/is steady.

The second part involved quantitative modeling. The index that we adopted was the Centa-City Leading Index (CCL). It is a weekly index based on the current contract prices in Centaline Property Agency Limited transactions (Centa Data 2013) (Centaline Property Agency is one the largest real estate agency companies in Hong Kong). It was constructed based on the preliminary contract price data: this often provides valuable information about recent property price movements. Since Centaline Property Agency Limited often have more than 20% of property agent's market transactions in the presence of real estate agents, it is generally believed that CCL is able to reflect the property market situation in Hong Kong. In short, CCL is: total market value of the constituent estates in a week/total market value of the constituent estates in the previous week x CCL for the previous week.

On top of that, we also collected data on the money supply (M1), interest rate, population, unemployment rate and GDP per capita. M1 rose from 277140.93 2003 to 948375.3 April 2013. Details of the macroeconomic data are shown in Table 5.1.

Table 5.1 Summary for the data used in the Hedonic Pricing model (Hong Kong Monetary Authority 2014, Census and Statistics Department 2014)

Year and month	M1	Interest rate (HK)	Population ('000)	Unemployment rate	GDP per capita in chain dollars	Writers are celebrity
2003 June	277140.93	1.1875	6 764.2	8.4	193867	17
2003 July	281982	1.28	6 764.2	8.4	186704	56
2003 August	285287.6	1.3438	6 764.2	8.2	186704	106
2003 September	299019	1.03	6 764.2	8.2	193867	100
2003 October	347044.5	1.2188	6 764.2	8.0	193867	17
2008 January	465464.9	2.225	6 963.9	3.3	254772	172
2008 February	461679.3	2.1569	6 963.9	3.3	254772	200
2008 March	457591.5	1.875	6 963.9	3.3	254772	199
2008 April	449981.5	2.485	6 963.9	3.3	245406	151
2013 January	944704.5	0.849918	7 219.7	3.5	281355	29
2013 February	952427.2	0.845461	7 219.7	3.5	281355	98
2013 March	945074.2	0.846893	7 219.7	3.5	281355	73
2013 April	948375.3	0.849249	7 219.7	3.5	281355	81

5.5.1 Data description

Table 5.2 shows the data of Oriental Daily and Apple Daily. The year 2008 recorded the largest number of positive news.

Table 5.2 Summary of Oriental Daily and Apple Daily data

Months	2003			2008			2013		
	Positive	Negative	Neutral	Positive	Negative	Neutral	Positive	Negative	Neutral
Oriental Daily									
6	4	0	0	8	3	0	4	1	0
7	10	12	2	20	4	1	13	4	1
8	21	12	10	24	9	1	6	9	5
9	22	1	1	10	12	1	11	10	0
10	3	1	2						
Apple Daily									
6	4	4	3	9	4	5	8	0	0
7	18	11	3	27	0	4	19	3	4
8	38	7	4	10	10	2	10	19	1
9	38	4	13	5	14	2	10	13	6
10	4	1	0						

5.6 Heteroscedasticity-and-autocorrelation-consistent and ordinary least square regression models

To test whether market sentiments affect property prices, we included heteroscedasticity-and-autocorrelation-consistent (HAC) and ordinary least square (OLS) regression models to test the impact.

OLS methods were used for Hedonic modeling and were usually applied in housing research. The major merit of OLS is that it can be applied to a large data set. Nevertheless, it is limited by its powerlessness to take spatial autocorrelation into account. For example, a geo-statistical approach which used Maximum Likelihood Estimation can account for spatial autocorrelation. Hence, only relatively small data sets (1,000 observations or less) can be applied, depending on the number of variables and the complexity of the model. As a result, large data sets must either use OLS or be hampered by the inadequate specified estimates of relatively low significance (Neill *et al.* 2007).

The Hedonic Pricing model was applied to study the impact of various factors which affect property prices; an example is the rapidly developed suburban residential market near the downtown area of Savannah, Georgia in the US (Beck *et al.* 2012). It was also used to examine the impact of neighborhood characteristics on residential values within San Juan and studied the outcomes and importance of various neighborhood effects on dwelling values (Díaz-Garayúa 2009).

Traditional OLS is usually denoted as:

$$y_t = x + x_t \partial + \varepsilon_t$$

According to Sul *et al.* (2005), HAC estimation is based on scalar regression model setting:

$$y_t = x + x'_t \partial + \varepsilon_t$$

Or it can be expressed as in demeaned form:

$$\ddot{y}_t = x + \ddot{x}'_t \partial + \ddot{\varepsilon}_t$$

where ∂ represents the coefficient value under a set of exogenous variables x_t where the tilde affix signifies demeaning. Robust tests of ∂ involves the use of LRV estimation of variates in the form of:

$$V_t = W_t \varepsilon_t$$

where W_t is a vector of covariates or instruments.

Nevertheless, as ε_t is unobserved, it is replaced by $\ddot{\varepsilon}_t$ which is constructed according to regression residuals.

The major merit of the HAC model is that there may be heteroscedastic autocorrelation consistent standard errors as it can help correct a significant degree of serial correlation automatically in the residuals in OLS (Kishor 2007).

5.7 Research results

The results of the multiple regression models are shown in Table 5.3. Population displayed a positive and significant relationship with housing prices. Three out of four models suggested that the housing prices have a significant relationship with unemployment rate. Shiller (2007) suggested that changes in attitudes or psychology affect home prices. Although changes in attitudes could not be measured, media and journalists' reports in newspapers can be used as a proxy. The results of the Hedonic Pricing models, however, showed that the correlation of positive/negative news and property prices was insignificant. This implied that market psychology had no significant impact on property prices. Market news did not aid housing price prediction. This suggested that the housing price movements were closer to Eugene Fama's theory that asset prices were not predictable, which is consistent with the results in Brown and Chau (1997), and indirectly rejects Robert Shiller's proposition that rises and falls in asset prices are guided by investors' psychology: it is hard to predict the ups and downs of housing prices by studying market investors' behaviors. However, whether housing prices are predictable in the long run still remain an unresolved issue.

5.8 Conclusion

Similar to many commodities and assets in our market, housing prices are determined by a bundle of factors. Macroeconomic factors such as financial crisis, interest rates, money supply, housing policy affect people's ability and willingness to pay for property. As previous studies suggest, the news indirectly reflects market psychology, and can be used as a proxy to study if market change would provide important information on the direction of housing prices. The results, however, showed that both good

Table 5.3 HAC and OLS results

Modeling method	Model 1 HAC standard errors, bandwidth 6 (Bartlett kernel)	Model 2 OLS	Model 3 OLS	Model 4 OLS
Constant	−1286.49 ***	−1136.49***	−1286.49***	−1385.58***
M3	0.3339	−2.79369e-06	−4.44079e-06	−6.41212e-06
Interest rate Hong Kong	0.000231***	−2.05415*	−2.34989**	
Positive news	−0.100653	−0.480893	−0.100653	−0.0362374
Negative news	0.747696	0.239626	0.747696	0.763995
Population	0.198669***	0.175318***	0.198669***	0.208704***
Unemployment	−1.58491***	−1.41250***	−1.58491***	0.0561904
Celebrity writer	0.334871	0.242756	0.334871	0.296543
GDP per capita				9.91823e-05*
CCL(−1)		0.0902939***		

* significant at 90% level; *** significant at 99% level

and bad news do not have significant impact on housing prices. The ups and downs in asset prices were not guided by the psychology of investors in the market. Other market fundamentals, such as unemployment rates, interest rate and population, may have more important implications on housing prices in Hong Kong.

5.9 References

Beck, J., J. Fralick and M.Toma (2012) The Housing Price Bubble in a Suburban Georgia Setting: Using the Hedonic Pricing Model in the New South. *Journal of Applied Business Research*, 28(4), 651–660.

Beck, J. D. and S. Bhagat (1997) Shareholder Litigation: Share Price Movements, News Releases, and Settlement Amounts. *Managerial and Decision Economics*, 18, 563.

Berry, M. and T. Dalton (2004) Housing Prices and Policy Dilemmas: a Peculiarly Australian Problem? *Urban Policy and Research*, 22, 69–91.

Brown, G. and K. W. Chau (1997). Excess returns in the Hong Kong commercial real estate market. *Journal of Real Estate Research*, 14(2), 91–105.

Campbell, J. Y. and R. J. Shiller (2001) Valuation Ratios and the Long-run Stock Market Outlook: An Update. *NBER Working Paper 8221*.

Census and Statistics Department (2014) Labour Force. http://www.censtatd.gov.hk/home.html

Centa Data (2013) *Centa-City Index Hong Kong's Definitive Price Indices*.

Chau, K. W., B. D. MacGregor and G. M. Schwann (2001) Price Discovery in the Hong Kong Real Estate Market. *Journal of Property Research*, 18, 187–216.

Chau, K. W., S. K. Wong, C. Y. Yiu, M. K. S. Tse and F. I. H. Pretorius (2010) Do Unexpected Land Auction Outcomes Bring New Information to the Real Estate Market? *Journal of Real Estate Finance Economics*, 40, 480–496.

Díaz-Garayúa, J. R. (2009) Neighborhood Characteristics and Housing Values Within the San Juan, MSA, Puerto Rico. *Southeastern Geographer*, 49(4), 376–393

Fama, E. F. (1970) Efficient Capital Markets: A Review of Theory and Empirical Work. *Journal of Finance*, 25, 383–417.

Fama, E. F. (1991) Efficient Capital Markets II. *Journal of Finance*, 46, 1575–1618.

Hassan, S. A. (2011) Modeling Asymmetric Volatility In Oil Prices. *Journal of Applied Business Research*, 27, 71–77.

Hong Kong Monetary Authority (2014) *Monthly Statistical Bulletin*.

Hossain, B. and E. Latif (2007) Determinants of Housing Price Volatility in Canada: A Dynamic Analysis. *Applied Economics,*, 1–11.

Hui, E. C. M. and I. Nga (2009) Price Discovery of Property Markets in Shenzhen and Hong Kong. *Construction Management and Economics*, 27, 1175–1196

Karantonis, A. and X. J. Ge (2007) An Empirical Study of the Determinants of Sydney's Dwelling Price. *Pacific Rim Property Research Journal*, 13, 493–509.

Kishor, N. K. (2007) Does Consumption Respond More to Housing Wealth than to Financial Market Wealth? If so, Why? *Journal of Real Estate Finance and Economics*, 35, 427–448.

Krishnamurti, C., G. G. Tian, M. Xu and G. Li (2013) No News Is Not Good News: Evidence from the Intra-day Return Volatility–Volume Relationship in Shanghai Stock Exchange. *Journal of the Asia Pacific Economy*, 18, 149–167.

Lai, L. W., K. W. Chau, D. C. W. Ho and V. Y. Y. Lin (2006). Impact of Political Incidents, Financial Crises, and Severe Acute Respiratory Syndrome on Hong Kong Property Buyers. *Environment and Planning B: Planning and Design*, 33(3), 413–433.

Lee, C. L. (2009) Housing Price Volatility and Its Determinants. *International Journal of Housing Markets and Analysis* 2, 293–308.

Li, R. Y. M. (2009) The Impact of Climate Change on Residential Transactions in Hong Kong. *The Built and Human Environment Review*, 2, 11–22.

Li, R. Y. M. (2012a) "Chow Test Analysis on Structural Change in New Zealand Housing Price During Global Subprime Financial Crisis." In *18th Annual Pacific Rim Real Estate Society Conference*. Adelaide.

Li, R. Y. M. (2012b) The Internalization of Environmental Externalities Affecting Dwellings: A Review of Court Cases in Hong Kong. *Economic Affairs*, 32, 81–87

Li, R. Y. M. (2013) The Usage of Automation System in Smart Home to Provide a Sustainable Indoor Environment: A Content Analysis in Web 1.0. *International Journal of Smart Home*, 7 47–59.

Li, R. Y. M. (2014) *Law, Economics and Finance of the Real Estate Market – A Perspective of Hong Kong and Singapore*. Germany: Springer.

Li, R. Y. M. and H. Y. Chan. (2013) "The Impact of Eurozone Debt Crisis on China's Property and Land Market." In *European Union Studies Association Asia Pacific Conference*. Macau.

Li, R. Y. M. and J. Li. (2012) "The Impact of Subprime Financial Crisis on Canada and United States Housing Market and United States Housing Cycle and Economy." In *ICBMG Conference*. Hong Kong.

Li, R. Y. M. and H. Y. Tsoi (2014) Latin America Sustainable Building Finance Knowledge Sharing. *Latin American Journal of Management for Sustainable Development*, 1, 213–228.

Li, R. Y. M. and T. T. Wong. (2013) "The Shadow of Eurozone Sovereign Debt Crisis Kept the Eastern Dragon's Property Away from Sunshine? Time Series Analysis on the Negative Externality in China's Property Market by EView 7 Software." In *International Symposium on Computational and Business Intelligence*. New Delhi, India.

Lucke, B. (2013) Testing the Technology Interpretation of News Shocks. *Applied Economics*, 45, 1–13.

McCollough, J. and K. Karani (2014) The Media's Role in Influencing Perceptions of Housing Values and the Resulting Impact on the Macroeconomy. *Economic Affairs*, 34, 68–77.

Milgrom, P. R. (1981) Good News and Bad News: Representation Theorems and Applications. *The Bell Journal of Economics*, 12, 380–391.

Neill, H. R., D. M. Hassenzahl and D. D. Assane (2007) Estimating the Effect of Air quality: Spatial Versus Traditional Hedonic Price Models. Southern Economic Journal, 73(4), 1088

Newell, G. and K. W. Chau (1996) Linkages Between Direct and Indirect Property Performance in Hong Kong. *Journal of Property Finance*, 7(4), 9–29.

Nguyen, T. (2011) US Macroeconomic News Spillover Effects on Vietnamese Stock Market Nguyen. *The Journal of Risk Finance*, 12, 389–399.

Nguyen, V. H. and E. Claus (2013) Good News, Bad News, Consumer Sentiment and Consumption Behavior. *Journal of Economic Psychology*, 39, 426–438.

Pope, J. C. (2008) Do Seller Disclosures Affect Property Values? Buyer Information and the Hedonic Model. *Land Economics*, 84, 551–572.

Sago, B. and C. Hinnenkamp (2013) The Impact of Significant Negative News on Consumer Behavior Towards Favorite Brands. *Global Journal of Business Research*, 8, 65–72.

Schwann, G. M. and K. W. Chau (2003) News Effects and Structural Shifts in Price Discovery in Hong Kong. *Journal of Real Estate Finance and Economics*, 27, 257–271.

Shiller, R. J. (1981) The Use of Volatility Measures in Assessing Market Efficiency. *Journal of Finance* 36, 291–304.

Shiller, R. J. (1984) Stock Prices and Social Dynamics. *Cowles Foundation Discussion Paper No.719.*

Shiller, R. J. (2003) From Efficient Markets Theory to Behavorial Finance. *Journal of Economic Perspectives,* 17, 83–104.

Shiller, R. J. (2006) Long-term Perspectives on the Current Boom in Home Prices. *Economists' Voice*, 3, 1–11.

Shiller, R. J. (2006) *Irrational Exuberance.* US: Crown Business.

Shiller, R. J. (2007) Historic Turning Points in Real Estate. *Cowles Foundation Discussion Paper No.1610.*

Soroka, S. N. (2006) Good News and Bad News: Asymmetric Responses to Economic Information. *Journal of Politics,* 68, 372–385.

Steeley, J. M. (2004) Stock Price Distributions and News: Evidence from Index Options. *Review of Quantitative Finance and Accounting,* 23, 229–250.

Sul, D., P. Phillips and C. Choi (2005) Prewhitening Bias in HAC Estimation. *Oxford Bulletin of Economics and Statistics,* 67, 517–546.

Xu, J. and J. J. Wang (2012) Reassembling the State in Urban China. *Asia Pacific Viewpoint,* 53, 7–20.

Yanga, Z. and B. Turner (2004) The Dynamics of Swedish National and Regional House Price Movement. *Urban Policy and Research,* 22, 49–58.

6 Superstition and Hong Kong housing prices

A Hedonic Pricing approach

Rita Yi Man Li, Kwong Wing Chau,
Chui Yee Law and Tat Ho Leung

6.1 Housing prices, transportation and superstition

Previous literature has shown that there is a bundle of attributes which affect property prices, such as environmental externalities (Li 2012b, Li and Li 2011), property management (Hui *et al.* 2011), weather conditions (rainfall and temperature (Li 2009)) and quality of life (Burinskienė *et al.* 2011). Li (2014)'s HP filter model showed that there was a close relationship between business cycle and housing prices. Narwold (2008)'s model showed that housing prices increase by 3.8% when there is a historical building within 250 feet and 1.6% between 250 and 500 feet. In Hong Kong, many of the housing attributes – apartment size, floor level, age – were not priced equally by homebuyers (Choy *et al.* 2012).

Geography is another factor which affects housing price movements. As some individuals may relocate to lower cost living areas, it is reasonable to observe the correlation between places in close proximity. For example, some people may choose to live in Shenzhen and work in Hong Kong, traveling across the border daily. Although the traveling costs are higher, the cost of living – such as dining and amenities – are a lot cheaper. That also implies that the high living costs in one place may spill over to the city nearby, as lots of us are quite mobile. And we may move from one city to another city. Moreover, the economic prosperity in one place may also bring some fortune to places nearby: the rich may also spend money on economic activities in the nearby city. And good economic performance often boosts the demand for better quality housing. With this in view, Li (2014)'s research shows that housing prices in the US and Canada shared similar ups and downs over a decade. The results of a Chow test analysis showed that the financial crisis had a spillover effect on property prices (Li 2012a).

In urban areas with railway transportation, many research articles shed light on the relationship between railway and property prices. In the log–linear Hedonic Pricing model, when distance from a railroad line doubles in Norway, property prices increase by about 10% within a 100 meters. The effects were even stronger when the distance was less than 100 meters from the train line (Strand and Vagnes 2001). Modernization of a railway line that connects the urban and suburban areas usually lead to a movement of the population from urban to suburban areas. This affects the residential price gradient of the two connected stations on the railway line. The price gradient may increase due to higher population density and the worsened living environment of suburban areas. Conversely, this may lead to a reduction in the pricing gradient between the suburban and urban areas due to an increase in demand for housing units in suburban areas. The time cost of commuting: an improvement in public transportation lowers the costs of commuting which results in a decrease in the difference in prices of properties along a railway line (Chau and Ng 1998).

Debrezion *et al.* (2003) pointed out that the transportation system had a positive effect on property prices. The meta-analysis evidenced that commercial properties had greater impact on price than residential area. Furthermore, the impact of commuter railway stations had a stronger impact on property price than heavy or light railway, or metro station. It is because the services provided by railway may have larger coverage, leading to a greater impact on property prices.

Apart from transportation and environmental externalities, there are a few studies which shed light on the lucky and unlucky factors which affect property prices. According to a sample of apartment transactions from 2004 to 2006 in Chengdu, China, the numbers "8" and "6" are lucky numbers, whereas "4" is unlucky. It was found that secondhand apartments located on the floors ending with "8" were 235 RMB higher priced per square meter than the other floors on average. Although this price premium disappears due to uniform pricing of new housing units for newly constructed apartments, apartments on floors ending in an "8" were sold 6.9 days faster on average than the other floors (Shum *et al.* 2014).

Furthermore, buyers may be unwilling to live in a residential unit which contained a murder or death by other unnatural cases. A sample of 4,893 apartment transactions in one of the most popular estates in Hong Kong showed that transaction prices are reduced by around 7.5% for stigmatized flats (Man and Wong 2012). People were willing to pay a premium for a "lucky" property. For example, "8" is a lucky number as eight has a similar pronunciation to "prosperity" in Cantonese. Therefore, a lucky floor number is often considered to be a valuable attribute. A Hedonic Price model showed that residential units with lucky floor numbers were sold at significantly higher prices during boom periods rather than in property slumps (Chau *et al.* 2001).

Past research has evidenced that there are differences between the generations. For example, the (young) generation Y relied more on government help on homeownership than the previous generation X (Li 2015).

Hence, we speculate that the younger generations may have different views with regards to lucky and unlucky numbers. Specifically, the previous generation are more superstitious than the present generation of homebuyers because there is a change in culture. For example, some young people are willing to move into residential units in which people died due to suicide or unnatural causes, while many in the older generation avoid such units, believed to bring bad luck. Previous research has shown that culture can inform human behavior. Accordingly, we think that the change in the impact of lucky numbers on property prices is mainly due to a change in culture (Li 2010, Li 2014, Li 2011). We put this down to three major reasons:

1   The level of education is usually higher (previous generation may not have received much education but the present generation receive at least nine years of education).
2   The influence of Western beliefs has made the present generation less superstitious.
3   Many of local news outlets report that there is a vast amount of young people who are not afraid to occupy public housing with previous records of unnatural deaths. This perspective on rent may be extended to private housing purchases in our present research.

Based on the above three reasons, we hereby propose that the impact of traditional lucky and unlucky numbers on relatively new districts, with younger families and households, is likely to be limited. As the older generations are aware of the adverse impact (mainly lower property prices or difficulty in resale) brought by bad numbers, many of the last decade's housing developments have already had all the so-called unlucky numbers, such as 4, 14, 24, removed. Accordingly, we used relatively new housing developments, which have not yet removed these numbers. That, however, also implies that the latter are not the most up-to-date buildings.

6.2 Data collection

We used the Hedonic Pricing model to examine the relationship between lucky and unlucky numbers on housing prices. In this chapter, private housing estates' property prices near the Po Lam Station are included. They include Metro City Phase 1, 2, 3, Verbena Heights, Well on Garden, Finery Park and Serenity Place. The scale of the housing developments varies from 6,768 housing units in Metro City to 688 housing units in Finery Park. The buildings' ages range from 13 to 19 years. Whilst Finery Park and Metro City were built by Henderson Land, Well On Garden was built by Nan Fung Group. All these housing estates are near Po Lam MTR station (MTR stations can be considered as a train stations).

All the transaction data was obtained from Centaline Property Agency Limited (2013) and macroeconomic data was obtained from Census and Statistics Department (2013). We have included dummy "lucky" floor numbers, which start or end with "8" and "3." Flat numbers beginning or ending with "4" were also included as a dummy unlucky number.

As property prices are affected by many factors, the study shall also include the housing attributes such as the distance of MTR and age, floor, gross area into the Hedonic model and macroeconomic factors such as inflation and unemployment are also included. The maximum housing price was about seven million and the minimum price was about one million with an average of around three million.

Table 6.1 Summary of the five housing developments in Tseung Kwan O

Estate name	Metro City	Verbena Heights	Well on Garden	Finery Park	Serenity Place
Developer	Henderson Land Development	Hong Kong Housing Society	Nan Fung Group	Henderson Land Development	Hong Kong Housing Society
Building age	17 (Phase1) 13 (Phase2) 11 (Phase3)	17–16	19	19	14
No. of blocks	21	6	4	2	5
Total no. of flats	6768	1894	1280	688	1526

Table 6.2 Statistics summary of the data which has been included in this chapter

	Maximum	Minimum	Mean	Standard deviation
Price	7,100,000	1,250,000	3,345,168.1	956,136.71
Age	19	11	15.7	2.38
Floor	48	1.0	21.2	12.14
Gross area	1026	456	639.5	128.58
Inflation	7.9	1.0	4.0	1.32
Unemployment	4.8	3.0	3.6	0.46
Distance	513.6	70.6	283.7	150.37

We also included the macro economic data of unemployment rate and inflation alongside with the housing data such as age, floor, gross area of the housing units and distance to the MTR station.

6.3 Hedonic Pricing model

The Hedonic Pricing model can be applied to explain the different values of hetero-geneous products and goods. It is often used as a technique to assess property value, estimate the demand for some of the specific attributes of housing, neighborhoods and analyze the price indexes for various types of properties. This heterogeneity is reflected in the different characteristics of housing. It studies the implicit price of each of the housing characteristics (Beekmans *et al.* 2014). It is used to estimate the marginal contribution of the individual characteristics which may affect the housing prices. The Hedonic model accepts that consumers attach certain values to some of the characteristics according to a derived utility for them, such that price equals to a function of these attributes (Fields *et al.* 2013). The Hedonic model usually includes property price as a dependent variable with a number of explana-tory or dependent variables such as the environmental factors, binary variables and lagged variables. As mentioned in Chapter 1, it is a very popular econometric method to study the factors which affect the housing price. The earliest application can be dated back to Rosen's research. He provided a basic framework for the Hedonic Pricing model (Rosen 1974):

$$P_i = f(S_i, N_i) \tag{1}$$

where P_i = property price i,
   S_i = vector of property's structural characteristics and
   N_i = vector of neighborhood characteristics

Kang and Reichert (1987) proposed that there was no single functional form without shortcoming. Thus, the functional form selection depends largely on whether the appraiser's goal is to minimize mean prediction error or to maximize prediction stability (Kang and Reichert 1987, Kryvobokov 2013). The Hedonic model includes the demand and supply of those attributes under different assumptions with regards to market structure, preferences and technology (Rosen 1974, Ekeland, Heckman and Nesheim 2002). In general, it is used to estimate the marginal contribution of those individual characteristics (Sirmans, Macpherson and Zietz 2005).

Cebula (2009) modified the dependent variable into logarithm function and binary explanatory variables. The Hedonic Pricing model had been used in the research studies of the Savannah historic landmark district from 2000 to 2005. The factors include interior and exterior physical characteristics for properties.[1] In southern California, a Hedonic Pricing model was used to study the short and long term impact of repeated wildfires on housing prices. On the other hand, Li (2009) used the Hedonic Pricing model to study the impact of various climate factors' impact on housing prices in Hong Kong.

The Hedonic model can be used to analyze the mortgage lending industry (Laurice and Bhattacharya 2005). It makes good use of a large, pooled cross-section of data to establish price prediction models in different regions[2] (Laurice and Bhattacharya 2005). In this chapter, we have used it to study the impact of lucky numbers and transportation on housing prices.

6.4 Research results of Hedonic Pricing models

According to the results, the adjusted R^2 is 0.699858, which implies that nearly 70% of the variations or changes in housing prices can be explained by this model. Most of the variables were statistically significant at 1%, with p-values smaller than 0.01 or 0.05. As an increased number of variables shall lead to an increase in the R^2 value, this will affect the percentage of variation that can be explained by the model, and the adjusted R-squared would be more reliable.

---

1 The model establishes as: $\ln P_i = f(I_j, E_j, SC_j, O_j)$.
2 The regions included Southern California counties, Los Angeles County, Orange County and San Diego County from January 2002 to June 2003.

Table 6.3 Variables included in this chapter (housing price is the dependent variable)

Independent variables	Descriptions
Age	The age of the residential units
Floor	The floor of the residential units
Dist	The shortest distance between Po Lam MTR Station and estates measured in meters.
Inflation	Inflation at the time of transaction
Unemployment	Unemployment at the time of transaction
Grossarea	The total area of the flat unit
Lucky_C	1 if the floor of the flat transacted is a lucky number in Chinese, zero if otherwise
Unlucky_C	1 if the floor of the flat transacted is an unlucky number in Chinese, zero if otherwise.

Table 6.4 Expected results of various variables

Variables	Expected result	Reason for the expected results
Age	Negative	Older flats are obsolete and less attractive
Floor	Positive	High floor level units have better views
Dist	Negative	The longer distance between home and MTR increase the commuting time
Inflation	Positive	Property prices increase with the inflation
Unemployment	Negative	High unemployment rate lowers buyers' affordability
Grossarea	Positive	Larger gross floor area provides spacious and comfortable living environment
Lucky_C	Positive	If the floor number is the lucky number, it brings good luck to the flat owner. People are willing to pay more.
Unlucky_C	Negative	If the floor is the unlucky number, it will bring bad fortune to the flat owner.

The model showed that the transaction price of the property was HK$ 3.664 million if the flat was completely new, no distance from the MTR, with no floors, zero gross floor area, no inflation, and a zero unemployment rate. The coefficients representing transaction price will increase 11949.64, –2038.593, 7009.264, 4831.706, 5302.168, 33057.93, –883936.4, –7595.646 if the age, distance, floor, gross floor area, inflation, luck, unemployment and lack of luck are one respectively.

Age, floor and gross floor area increases with the transaction price significantly at the 1% level while distance and unemployment decreases the transaction price significantly at that level. The unlucky/lucky number relationship with property prices is positive but insignificant, implying that Hong Kong people may not be superstitious; rather, they prefer a higher floor with a better view. The age of the estate is positive and significant; ages varied from 19 to 11. Older estates have established reputations, so homebuyers have more information about them.

Table 6.5 Results of the Hedonic Pricing model

Variables	Coefficient	P-value
Intercept	3664258***	0.0000
Age	11949.64**	0.0153
Dist	−2038.593***	0.0000
Floor	7009.264***	0.0000
Grossarea	4831.706***	0.0000
Inflation	5302.168	0.5950
Lucky_C	33057.93	0.2545
Unemployment	−883936.4***	0.0000
Unlucky_C	7595.646	0.8051
R^2	0.699858	
Adjusted R-squared	0.698843	

Notes

1 Coefficients statistically significant at 10%, 5% and 1% level are represented as *, ** and *** respectively.

2 P-values are a two-tailed test.

The results of this chapter showed that a one-meter increase in distance would decrease the property price by 2038.593, other factors being constant. In our research study, Metro City Phase II has the shortest distance from Po Lam MTR station while Serenity Place has the longest. That means, these two estates have a price difference because of the distance to the train station. This can be explained by the Hong Kong lifestyle, which puts a premium on convenience and efficiency. Time saved getting to the MTR was reflected in the property price. Therefore, it is not hard to imagine estate developers choosing to establish new estates near MTR stations at a higher unit price.

6.5 Limitation of the research study

In this chapter, we have only included some factors which might affect the property price, such as pollution, size and facilities. Moreover, each flat unit is heterogeneous even though the residential units are in the same estate, same building and floor as they may have different views and internal decorations. It is difficult to consider all the units as homogeneous. This implies that the model can only provide a rough estimation of the housing prices as there are too many factors in reality which could not be included, due to the high cost of data collection. On the other hand, previous research has revealed that where there are spatial auto correlations problems, the inclusion of various spatial econometrics techniques may capture some of these problems.

6.6 Conclusion

This research covered 2,380 transactions on five housing estates in Po Lam: Metro City Phase 1 to 3, Verbena Heights, Well on Garden, Finery Park and Serenity

Place during 2010 to 2013. In the model, age of the building, floor, distance from Po Lam MTR Station, gross area, inflation, unemployment and lucky number of the floor are included. The results show that an increase in distance from Po Lam Station leads to a decrease in the property price; closer properties tend to be more valuable. Therefore, Metro City Phase 2 has the shortest distance with the Po Lam Station and has a higher transaction price in general, other factors being constant. Transportation brings benefits of accessibility and promotes economic growth and convenience. These kinds of benefit are captured in the property price.

Nevertheless, the government and estate developers should realize that not every district in Hong Kong is suitable for an MTR station. Some districts, such as Sai Kung, are characterized by silence, low density and distance from central business districts. Home owners in these districts usually drive to work, making the MTR redundant.

Furthermore, though research suggests that lucky numbers increase the price of property, and vice versa, such factors are predictably insignificant in this district. That confirms the authors' speculation that changes in informal institutions, i.e. culture, plays an important role. Many residents in Tseung Kwan O are quite young and educated. They are not superstitious about the good or bad luck associated with numbers. Accessibility, however, which affects their time to and from work, plays a more important role.

Last but not least, there are always some intrinsic and extrinsic amenities which may have significant influences on housing price determination but have not been included in our model. For example, one major drawback of our results is due to an unexplained variance in house prices relating to space (Dubé and Legros 2014).

6.7 References

Beekmans, J., P. Beckers, E. V. D. Krabben and K. Martens (2014) A Hedonic Price Analysis of the Value of Industrial Sites. *Journal of Property Research*, 31:2, 108–130.

Burinskienė, M., V. Rudzkienė and J. Venckauskaitė (2011) Effects of Quality of Life on the Price of Real Estate in Vilnius City. *International Journal of Strategic Property Management,* 15, 295–231.

Cebula, R. J. (2009) The Hedonic Pricing Model Applied to the Housing Market of the City of Savannah and its Savannah Historic Landmark District. *The Review of Regional Studies,* 39, 9–22.

Census and Statistics Department (2013) *Browse by Subject*.

Centaline Property Agency Limited (2013) *16 Years Transaction Records*.

Chau, K. W. and F. F. Ng (1998) The Effects of Improvement in Public Transportation Capacity on Residential Price Gradient in Hong Kong. *Journal of Property Valuation and Investment*, 16(4), 397–410.

Chau, K. W., V. S. M. Ma and D. C. W. Ho (2001) The Pricing of "Luckiness" in the Apartment Market. *Journal of Real Estate Literature*, 9, 29–40.

Choy, L. H. T., W. K. O. Ho and S. W. K. Mak (2012) Housing Attributes and Hong Kong Real Estate Prices: A Quantile Regression Analysis. *Construction Management and Economics,* 30, 359–366.

Debrezion, G., E. Pels and P. Rietveld (2003) *The Impact of Railway Stations on Residential and Commercial Property Value*. The Netherlands: Tinbergen Institute.

Dubé, J. and D. Legros (2014) Spatial Econometrics and the Hedonic Pricing Model: What about the Temporal Dimension? *Journal of Property Research*, 31, 333–359.

Ekeland, I., J. J. Heckman and L. Nesheim (2002) Identifying Hedonic Models. *American Economic Review*, 92, 304–309.

Fields, T. J., C. Earhart, T. Liu and H. Campbell (2013) A Hedonic Model for Off-Campus Student Housing: The Value of Proximity to Campus. *Housing and Society*, 40, 39–58.

Hui, E. C. M., H. T. Lau and T. H. Khan (2011) Effect of Property Management on Property Price: A Case Study in HK. *Facilities*, 29, 459–471.

Kang, H. B. and A. K. Reichert (1987) An Evaluation of Alternative Estimation Techniques and Functional Forms in Developing Statistical Appraisal Models. *Journal of Real Estate Research*, 2, 1–27.

Kryvobokov, M. (2013) Hedonic Price Model: Defining Neighborhoods with Thiessen Polygons. *International Journal of Housing Markets and Analysis*, 6, 79–97.

Laurice, J. and R. Bhattacharya (2005) Prediction Performance of a Hedonic Pricing Model for Housing. *The Appraisal Journal*, 73, 198–209.

Li, R. Y. M. (2009) The Impact of Climate Change on Residential Transactions in Hong Kong. *The Built and Human Environment Review*, 2, 11–22.

Li, R. Y. M. (2010) A Study on the Impact of Culture, Economic, History and Legal Systems Which Affect the Provisions of Fittings by Residential Developers in Boston, Hong Kong and Nanjing. *International Journal of Global Business and Management Research*, 1, 131–141.

Li, R. Y. M. (2011) *Everyday Life Application of Neo-institutional Economics*. Germany: Lambert Academic Publishing.

Li, R. Y. M. (2012a) "Chow Test Analysis on Structural Change in New Zealand Housing Price During Global Subprime Financial Crisis." In *18th Annual Pacific Rim Real Estate Society Conference*. Adelaide.

Li, R. Y. M. (2012b) The Internalization of Environmental Externalities Affecting Dwellings: A Review of Court Cases in Hong Kong. *Economic Affairs*, 32, 81–87.

Li, R. Y. M. (2014) *Law, Economics and Finance of the Real Estate Market – A Perspective of Hong Kong and Singapore*. Germany: Springer.

Li, R. Y. M. (2015) Generation X and Y's Demand for Homeownership in Hong Kong. *Pacific Rim Real Estate Journal*, 21(1), 15–36.

Li, R. Y. M. and Y. L. Li (2011) Judges' View on the Price of Environmental Externalities in the United Kingdom. *US–China Law Review*, 8, 994–1007.

Man, K. F. and V. Wong (2012) Haunted Flats: Quantifying the Value of Stigmatization in an Apartment Market. *Appraisal Journal*, 80, 330–336.

Narwold, A. J. (2008) Estimating the Value of the Historical Designation Externality. *International Journal of Housing Markets and Analysis*, 1, 288–295.

Rosen, S. (1974) Hedonic Prices and Implicit Markets: Product Differentiation in Pure Competition. *Journal of Political Economy*, 82, 34–55.

Shum, M., W. Sun and G. Ye (2014) Superstition and "Lucky" Apartments: Evidence from Transaction-level Data. *Journal of Comparative Economics*, 42, 109–117.

Sirmans, G. S., D. A. Macpherson and E. N. Zietz (2005) The Composition of Hedonic Pricing. *Journal of Real Estate Literature*, 13, 3–43.

Strand, J. and M. Vagnes (2001) The Relationship between Property Values and Railroad Proximity: a Study Based on Hedonic Prices and Real Estate Brokers' Appraisals. *Transportation*, 28, 137–156.

7 Negative environmental externalities and housing price

A Hedonic model approach

*Rita Yi Man Li, Kwong Wing Chau,
Ming Hong Li, Tat Ho Leung and
Ka Wai Kwok*

7.1 Introduction

Similar to any other commodities, housing prices are driven by a number of different factors. Previous literature showed that structural and location factors affect housing prices (Tekel and Akbarishahabi 2013). In Chinese society, Feng Shui is believed to be an important factor which may drag housing prices up or down. The word "Feng Shui" is equivalent to the word wind and water in English. It is the desire to bring good fortune to our offspring by leaving better land (Li, Law and Leung 2014, Chau *et al.* 2001). It is believed that good Feng Shui benefits households while bad Feng Shui brings bad luck to residents. Moreover, many Chinese believe that some a properties' characteristics and views can bring different fates to the residents who live there. For example, it is generally believed that a sea view brings wealth to the household: sea represents water and water is related to money and riches.

Furthermore, environmental factors such as air pollution, noise pollution and water pollution affect the quality of environment and health. For example, poor air quality in Beijing has led to an increase in hospital admission rates due to asthma and other health problems. Hence, noise from railways or airports, water and air pollution, for example, have a negative effect on property prices (Anstine 2003, David 2006, Nelson 2004). Anstine (2003) examined the impact of two factories near a property in a semi-rural area. It was assumed these noxious facilities would affect house values significantly. However, if the information cost is so high that residents near the polluted area fail to receive information through the usual senses – such as sight and smell – the negative value caused by environmental pollutants might not be affect housing values.

Nelson (2004) applied meta-analysis to study the impact of noise pollution on residential property values in the US and Canada. The weighted-mean noise discount was 0.58% per decibel, covered 33 estimates, including 23 airports in the US and Canada. Furthermore, the cumulative noise discount in the United States was about 0.5% to 0.5% per decibel at noise exposure levels of 75 dB or less while the discount was 0.8% to 0.9% per decibel in Canada. It was also found that homebuyers are willing to pay a lot more for a one degree celsius rise in temperature (Li 2009).

Furthermore, accessibility of public facilities can boost property prices. These include parks and footpaths in foreign countries, public transport, restaurants and shopping malls in Hong Kong (Limsombunchai *et al.* 2004b, Cebula 2009a, Kong *et al.* 2007). Different functional forms of the Hedonic Pricing model, such as logarithms or squaring, were applied to study the impact of these facilities on housing prices (Cebula 2009a, Laurice and Bhattacharya 2005b, Tekel and Akbarishahabi 2013).

A large property can provide a spacious environment: property size usually displays a positive relationship with price. Concerning the property age, it usually has negative relationship with property price. However, there are some occasions that evidence positive relationship. Cebula (2009a)'s results showed that property price can be positively correlated with the property age if that property has been significantly designated as a National Historic Landmark.

Sirmans *et al.* (2005a) found that the number of rooms is also an important factor. Different types of rooms (bedroom, kitchen and bathroom) had different levels of impact on housing prices. Some research estimated the effect of a half or full bathroom. Previous research found that an increase in the number of bedrooms had a positive correlation with price. Meanwhile, a half bathroom had much less of an impact on a property's price than a full bathroom; though both had a positive effect.

On top of that, the internal features of properties might also affect price: some of us are attracted by the appearance of housing units. Culture values often affect homebuyers' decision making. For example, there is another factor that could contribute more in Hong Kong than in other countries – what floor the property is on (Sirmans *et al.* 2005a). Kryvobokov (2013) estimated the relationship between floor level and price. Kryvobokov showed that higher floors are often associated with higher prices. Similar results can also be observed in Hong Kong: housing units on lower floors are usually tied in with poorer views and a higher chance of foul water leakage. Not in the same vein, however, some Virginia residents in the US reported that they disliked living in residential units on higher floors. This implies that those on the lower floors may be sold at higher prices.

Property practitioners often put "location, location, location" on their lips. Indeed, location is a critical factor which affects property prices. Various neighborhood factors – the crime rate, education, average income, proximity to historic landmarks –often play important roles in determining housing prices. Location includes other factors, such as distance to a shopping mall, leisure in the countryside or zoning in urban, rural, industrial or popular school zones (Cebula 2009b).

Hence, we may observe that there are many variables which are related to neighborhood and location, it is unlikely that each factor's effect is the same. Some might have more adverse impacts than others. Indeed, some of the factors have stronger correlation with property price. Goodman (1978a) stated that places with more black and poor people should also be included in Hedonic Price model analysis. Sirmans *et al.* (2005a) suggested that crime rate should also be included. For example, some cities with more certain types of race are also associated with higher crime rates. In Australia, suburbs with more aboriginals may have lower property prices due to the possibility of higher crime rates. On the other hand, a popular school district in Hong Kong, such as Kowloon Tong, attracts many parents, leading to a stronger demand for housing and relatively higher prices. In view of this, it is quite natural to observe some researchers include schooling as one of the variables in the Hedonic Pricing model.

Transaction-related factors include: the quality of the properties sold, the owner's occupation, time trend, the month of the transaction and the holding period before sale. The property quality at sale affects the selling price. If it is poor, the number of buyers who are willing to buy will drop. Moreover, a quality property can command a high price. Laurice and Bhattacharya (2005b) estimated the trend of the market to determine the seasonal impact on housing. Witte *et al.* (1979) focused on consumer factors, such as: supplier characteristics (annual income), owner occupation (blue collar or white collar), education level, length of ownership, ownership form and the owner's race as supplier factors to price model.

Sirmans *et al.* (2005a) indicated that tax and financial factors had negative or no impact on property prices. In Hong Kong, the property tax is not high. Nevertheless, the effect of a property tax increase cannot be neglected as that might constitute a huge cost on home purchase. Investors and speculators around the world have a large incentive to invest in the Hong Kong real estate market. Equally, property owners, as investors in Hong Kong, have their own expectations of the real estate market. For example, they tend to hold on to their property rather than put it on the market if the expectation is positive in the following months.

As a matter of fact, noise pollution adversely affects our daily activities or health. Amoy garden in Kowloon Bay was selected since it is one of the largest property developments in Hong Kong with a number of housing towers that face the main road: the Ngau Tau Kok Road has four lanes and there are heavy traffic flows every day. It was expected there would be serious noise generated by those motor vehicles. High noise levels can contribute to heart disease in humans, causing an increase in blood pressure, stress and likelihood of heart disease. The price of residential units facing the road was expected to be lower than the others since most of us dislike noise. This chapter studied the impact of externalities' impact on property values in Hong Kong from 1994 to July 2013. It mainly shed light on the effect of noise pollution and air pollution on property values.

7.2 Research method

Due to the differences of residents' preference, there is heterogeneity of hedonic prices (Goodman and Thibodeau 2003). Although it was widely used, there were some difficulties which might affect the accuracy of the model. For example, Pennington *et al.* (1990) pointed out that there were challenges to predict the effect of aircraft noise on residential property values when this model was applied in their research study.

Many of the scholars tried to improve the model by using different statistical methods. For example, Can (1990) used Moran's I to test for autocorrelation and Bowen *et al.* (2001) advocated the use of spatial diagnostics in hedonic house price estimation. Various econometric techniques, such as maximum likelihood

Table 7.1 List of the factors which may affect property prices and can be included in the Hedonic Pricing model

Category	Factors	Detail	Expected relationship
Construction and structure	Property size		+
	Property age		−
	Number of rooms	Bedrooms, half bathrooms, full bathrooms . . .	+
	*Windowsills		−
Internal features	Fireplace		+
	Air-condition system		+
	Basement		+
	Floor	The floor of property located	+
External features	Garage size		+
	Carport		+
	Courtyard		+
	Pool		+
	Porch		+
	*Clubhouse	Clubhouse entertainment room and pool	+
Environment	Good landscape[a]		+
	Ocean-view		+
	*Mountain-at-the-back		+
	*Feng Shui		+
Environmental	Noise pollution		−
	Water pollution		−
	Air pollution		−
Surrounding facilities	Facilities for leisure entertainment	Walking path, park	+
	Facilities for convenience life	Restaurant, shopping mall, public transport, convenience store	+
Neighborhood and location	Crime rate		−
	Majority race	Percentage of minority race group in that country	−
	*School zone		+
	Type of location	Rural, industrial, urban . . .	?
	Distance	Distance to city center, distance to countryside	−
	Danger zone	Earthquake, flood	−
Transaction related	Property quality	Need decoration	?
	Time trends		?
	Transaction month		?
Financial issues	Property tax		?
	Future expectation		?

Sources: Cebula 2009b, Goodman 1978b, Laurice and Bhattacharya 2005a, Limsombunchai *et al.* 2004a, Sirmans *et al.* 2005b, Li 2014, Li and Li 2011, Li 2012.

a Although good landscape is a relatively subjective concept, Sirmans, G.S., Macpherson, D.A. & Zietz, E.N. (2005) used this as a variable for estimation.

* refers to the factors which should fit Hong Kong's case.

(Brassington 1999), mixed regressive-autoregressive (Simons 2001), Q-factor analysis (Dale-Johnson 1982), principal component analysis (PCA) (Maclennan and Tu 1996, Bourassa *et al.* 2003) have been used to adjust spatial autocorrelation. However, a popular and commonly used method is Rosen's Hedonic model (Rosen 1974).

Rosen's hedonic equation "represents a joint envelope of a family of value functions and another family of offer functions" (Rosen 1974). Therefore, the equation contains two sides: consumers (demand) and suppliers (supply). With regards to the function of the consumers, Rosen defines a value or bid function for each of the households which indicates the maximum amount that the consumers are willing to pay for the alternative housing bundles, C's. These housing bundles contain different amount of attribute ($C = c_1, c_2, c_3 \ldots\ldots c_n$) which indicates the consumers' utility. In addition to the features of the housing bundle, the consumers' bid, β, for a given bundle is affected also by their income level (y) and tastes (π). Thus, the formula for housing consumers is:

$$\beta = \beta(c_1, c_2, \ldots, c_n, y, \pi)$$

where π is a vector of taste. For function of suppliers, Ω, represents the minimum unit price the producers are willing to accept for the housing. Assuming that firms are rationally and profit maximized, the design or constituent of the function depends on the characteristics of the bundle ($C = c_1, c_2, c_3 \ldots\ldots c_n$), factor prices (f) and production parameters (p) if it is sold in the first market. Therefore, the formula is: $\Omega = \Omega(c_1, c_2, \ldots, c_n, f, p)$ where f and p is a vector that contains various factor prices and production parameters. In market equilibrium, it is attained when $Q^S(C) = Q^D(C)$, $Q^S(c_i) = Q^D(c_i)$ or that β = Ω, for all i.

7.3 Data description

To estimate the impact of various factors on property prices, we have included three major groups of data in our study:

1 Various types of housing attributes such as area, floor level of the housing units.
   This group of data was obtained from Centaline Property Agency (2014).

2 The macro economic factors such as exchange rate between RMB and daily Hong Kong dollar, Consumer Price Index (CPI), GDP.
   The exchange rate data was obtained from Yahoo Finance, CPI and GDP data was obtained from Hong Kong Census and Statistics Department.

3 Environmental factors such as Sulphur dioxide, nitrogen dioxide, oxone, respirable particulates.
   This dataset is obtained from Hong Kong Environmental Protection Department.

In short, Table 7.2 summarizes the data used in this chapter:

Table 7.2 Data description table

Variable names	Description	Type	Data source
Price	Transaction price (in million)	–	Hong Kong CENTADATA[a]
Area	Area of the flat (in ft^2)	–	Hong Kong CENTADATA[b]
Ex	Exchange rate between RMB and HKD $= \dfrac{CNY}{HKD}$	Daily	Yahoo! Hong Kong -Finance[c]
GDP	Hong Kong GDP at current market prices (in million)	Quarterly	Hong Kong Cenus and Statistics Department[d]
CPI	Composite Consumer Price Index	Monthly	Hong Kong Cenus and Statistics Department[e]
FLOOR_ DUMMY	Flats in the lowest ten floors: FLOOR_DUMMY = 1, otherwise = 0.	–	Hong Kong Centadata
ROAD_ DUMMY	Flats facing Ngau Tau Kok Road ROAD_DUMMY = 1, otherwise = 0	–	Hong Kong Centadata
SO2	Sulphur Dioxide (in μg/m3)	Daily	Hong Kong Environmental Protection Department[f]
NO2	Nitrogen Dioxide (in μg/m3)	Daily	Hong Kong Environmental Protection Department
NOX	Nitrogen Oxides (in μg/m3)	Daily	Hong Kong Environmental Protection Department
O3	Ozone (in μg/m3)	Daily	Hong Kong Environmental Protection Department
RSP	Respirable Suspended Particulates (in μg/m3)	Daily	Hong Kong Environmental Protection Department

a Hong Kong CENTADATA
   http://hk.centadata.com/tfs_centadata/Pih2Sln/TransactionHistory.aspx?type=3&code=EWKSBP
   YOPS&info=basicinfo
b Hong Kong CENTADATA
   http://hk.centadata.com/tfs_centadata/Pih2Sln/TransactionHistory.aspx?type=3&code=EWKSBP
   YOPS&info=basicinfo
c Yahoo! Hong Kong -Finance http://hk.finance.yahoo.com/q?s=CNYHKD=X
d Hong Kong Cenus and Statistics Department Table 030 http://www.censtatd.gov.hk/showtable
   excel2.jsp?tableID=030&charsetID=2
e Hong Kong Cenus and Statistics Department Table 052 http://www.censtatd.gov.hk/showtable
   excel2.jsp?tableID=052
f Hong Kong Environmental Protection Department http://epic.epd.gov.hk/EPICDI/air/station/?lang=zh

　　Table 7.3 describes the mean, standard deviations and coefficients of deviation variation for housing price, area, GDP, CPI and pollutants level data. The average housing prices were 2.5 million dollars with an area of over 470 square meters. The exchange rate between RMB and the Hong Kong dollar was about 1.18.

Table 7.3 Descriptions of the factors which were included in the Hedonic Pricing model

Variables	Mean (million dollars)	Standard deviation	Coefficient of deviation variation
Price	2.525519759	0.585242542	0.231731524
Area	470.1334	47.83749	0.101753
Ex	1.183648189	0.039496307	0.033368282
GDP	469382.8	38401.73	0.081813
CPI	104.6597146	4.614529679	0.044090792
FLOOR_DUMMY	0.355104281	0.478676288	1.347987939
ROAD_DUMMY	0.061470911	0.240258028	3.90848327
SO2	10.30016	5.055576	0.490825
NO2	60.74625	20.5026	0.337512
NOX	118.2945	49.71141	0.420234
O3	34.62436	23.96145	0.69204
RSP	46.01813	22.53059	0.48960

7.3.1 Air pollutants data recorded in the Hong Kong Special Administrative Region (HKSAR) environmental protection department

The air pollution data, recorded by Kwun Tong Monitoring Station, was collected from HKSAR Environmental Protection Department. The location of the monitoring station is "Yue Wah Mansion, 407–431 Kwun Tong Road, Kwun Tong, Kowloon." It is 25 meters high. It is in an urban land use zone (HKEPD – Kwun Tong Monitoring Station 2013). As Kwun Tong Monitoring Station is close to Amoy Garden, it can provide a good estimate on the pollution problem in this housing estate. The distance between the Kwun Tong Monitoring Station and Amoy Garden is about two kilometers. Since the Environmental Protection Department do not provide daily pollutant data, the daily data used in the model is calculated by the means of 24 hours' hourly data.

$$\bar{x} = \frac{x_1 + x_2 + \cdots + x_{24}}{24}$$

x_1 = Hourly pollutants' data.
$\bar{x}$ = Daily pollutants' data.

7.3.2 Property transaction data in Amoy Garden

All the property transactions data (from January 2010 to June 2013) was collected from Centadata (Centaline Property Agency 2013). The floor, area and transaction price of flats were included in the dataset. There were 1,822 transactions (n = 1822) in the data set. Amoy Garden is a relatively large residential property development. It is also known by everybody in Hong Kong. This residential development was the first to have a serious SARS outbreak in 2003. Many residents died from SARS at that time. It is, therefore, expected that the residents may have a stronger desire for a positive environment than the other large scale housing developments.

There are 19 towers in Amoy Garden. They were built between 1980 and 1987. There are 33 floors in the tallest block and the shortest blocks have 26 floors. We chose it because high density often results in large housing transactions. As there are more transactions, there will be more observations and the results shall be more accurate. The transaction number of Amoy Gardens was large enough to allow a sufficient sample size for scientific research.

7.3.3 Data analysis

To improve the result of the models, some outliers were deleted from the data set. To find out the effect of facing traffic, a dummy variable was included. If the flats were facing a road, the dummy variable would be one or else zero. In the model, Flats 1 and 2 of Blocks A, B, I, J and K were defined as close to a road. A dummy variable (ROAD_DUMMY) equal to one was added to these flats. Since the transaction price, GDP and the exchange rate have a unit root, we replace them with their first difference.

7.4 The research results of four econometric models

Model 1 mainly focused on the effect of air pollutants on property values. As the coefficient of SO2, NO2, O3 and RSP are not significant at all, we could not conclude that the property values were affected by the air pollutants. On top of that, this model had low adjusted R^2. It means regressors X had low explanatory power on Y. But this model showed that area and floor of the residential unit would affect the change in property value. There was a positive relationship: if the area was larger, the property value may be higher. On the other hand, the transaction prices decreased by 0.17 million if the flats were on the lowest ten floors.

Model 2 focused on the impact of lower floor and air pollutants on property price. They helped us to study the interaction effect between flats on lower floors and air pollutants. The interaction between SO2, NO2, O3 and floor dummy was negative. It showed that lower floors' property values would decrease more than those on higher floors if SO2, NO2 and O3 increased. It might be because air pollutants mainly appear in lower heights of urban areas. The higher flats were affected by air pollutants less since the stronger wind improved the air quality.

In model 3, a floor dummy was added, which prompted a significant change. The variable NO2*FLOOR_DUMMY became positive and insignificant. SO2* FLOOR_DUMMY became more significant and the coefficient is negative as model 2. O3*FLOOR_DUMMY became less significant. It was significant at 90% confidence level only. And the coefficient was negative, which was the same as model 2. The adjusted R square of this model was the largest.

The adjusted R square was 0.515626. It meant the model explain 51.56% of Y (change in property price).

Model 4 excluded some of the insignificant variables based on the previous models. As the previous models showed that NOX and NO2 were always insignificant, model 4 had dropped these two variables. It showed that the floor of flats and whether it faces a road or not do affect property values. The coefficients of SO2 and O3 were out of our expectation. The coefficients were positive, which means that the relationship between air pollutants and property value was positive; the coefficients were significant at 90% confidence level. The result of RSP was as expected; it had a negative effect on property value at 95% confidence level.

With regards to the interaction variables, it was found that SO2* FLOOR_ DUMMY and O3*FLOOR_DUMMY were negative as expected. As mentioned before, compared to higher floors, the negative effect of air pollutants on lower floors was expected to be greater. Hence, all of the interaction variables are expected to have negative coefficient.

We found that three variables which include Area, FLOOR_DUMMY and ROAD_DUMMY were significant in all the above-mentioned four models. The floor area of the housing units displayed a positively significant relationship with the property values and flats located on the lower floors had lower prices. These

Table 7.4 Results of Hedonic Pricing model

Dependent variable: d (price)

Regressor	(1)	(2)	(3)	(4)
D (price)(-1)		−0.4922***	−0.49369***	−0.49349***
Area	0.004769***	0.004757***	0.004778***	0.004737***
D (ex)	2.22737	0.27596	0.209142	
D (gdp)	−1.76E−06	−8.93E−07	−1.02E−06	
FLOOR_DUMMY	−0.169833***		−0.19321***	−0.16005***
ROAD_DUMMY	0.067334*	0.089678***	0.089188***	0.088556***
SO2	−0.000587			0.004022*
NO2	0.000224			
NOX	0.000127			
O3	0.000259			0.000815*
RSP	−0.000365			−0.00148**
SO2* FLOOR_DUMMY		−0.00732*	−0.00801**	−0.00839**
NO2*FLOOR_DUMMY		−0.00133*	0.001257	
O3*FLOOR_DUMMY		−0.00218***	−0.00122*	−0.00201**
RSP*FLOOR_DUMMY		0.001833*	0.001545	0.00323***
D (ex)(-1)		5.493084**	5.255006*	
D (gdp)(-1)		1.76E−07	1.22E−07	
Intercept	−2.2002***	−2.1894***	−2.19048***	−2.17377***
Adjusted R^2	0.268884	0.510888	0.515626	0.512707

Note: *** is at 1% level, ** is at 5% level, * is at 10% level.

two variables were expected and consistent with the previous literatures. On the other hand, the ROAD_DUMMY variable (with the housing units facing the road side) had positive coefficients in all the econometric models which contradict with our expectation. As mentioned in the introduction, the price of flats face the road were expected to be lower than others since people dislike noise and air pollution. Nevertheless, the positive sign implied that there must be at least one variable which overrides the negative externalities due to pollution. We speculate that is due to the better view for those units which face the road.

7.5 Discussion and conclusion

The models showed that flats facing a road command a higher price. To find out why ROAD_DUMMY has a positive coefficient, some of the possible positive externalities factors should be examined. As the blocks of residential towers in Amoy Gardens are fairly close, some of the housing units only have views of blocks of towers. Those flats which face the road are not blocked by nearby buildings; many enjoy a view of Victoria Harbor. The view is spectacular: occupants can see fireworks on some occasions and buildings far away from their flats. The view of flats on higher floors is even better than those on lower floors. It offers a vivid explanation of why the dummy variable of lower floor in the models was negative.

Compared to flats facing a road, other flats might have worse views. Since the blocks in Amoy Gardens are close, views are easily blocked by other residential towers. In general, flats facing a road were more valuable than other flats as people were willing to pay more for a better view.

To conclude, this chapter has mainly studied the effect of noise pollution and air pollution on property value. Amoy Gardens was selected to be the study site. Different models showed that flats on lower floors had lower prices, while flats facing the road had higher prices. Moreover, property values of flats in lower floors decreased more than those on higher floors if air pollution was worse. The reason flats facing a road had higher property prices might be due to the positive externality of a good view. When the positive effect of a view was greater than the negative effect of air and noise pollution, flats facing a road would have a higher price.

Acknowledgement

An earlier version of this chapter was presented in Asian Real Estate Society.

Appendix

Model 1

Dependent Variable: TRAN_D
Method: Least Squares
Sample (adjusted): 2 1822
Included observations: 1718 after adjustments

Table 7.5 Results of Hedonic Pricing model 1

Variable	Coefficient	Std. error	t-Statistic	Prob.
C	−2.200197	0.101111	−21.7602	0
AREA	0.004769	0.000199	23.93732	0
EX_D	2.22737	3.371313	0.660683	0.5089
GDP_D	−1.76E−06	3.51E−06	−0.50047	0.6168
ROAD_DUMMY_1	0.067334	0.040056	1.680975	0.093
FLOOR_DUMMY_ 1_LOW_0_HIGH	−0.169833	0.020146	−8.43001	0
SO2	−0.000587	0.002816	−0.20847	0.8349
NO2	0.000224	0.001085	0.206273	0.8366
NOX	0.000127	0.000443	0.287866	0.7735
O3	0.000259	0.000672	0.384686	0.7005
RSP	−0.000365	0.000696	−0.52441	0.6001
R-squared	0.273142	Mean dependent var		0.000729
Adjusted R-squared	0.268884	S.D. dependent var		0.463407
S.E. of regression	0.396237	Akaike info criterion		0.992775
Sum squared resid	268.0058	Schwarz criterion		1.027664
Log likelihood	−841.794	Hannan-Quinn criterion		1.005684
F-statistic	64.14646	Durbin-Watson stat		2.598817
Prob (F-statistic)	0			

Model 2

Dependent Variable: TRAN_D
Method: Least Squares
Sample (adjusted): 3 1822
Included observations: 1716 after adjustments

Table 7.6 Results of Hedonic Pricing model 2

Variable	Coefficient	Std. error	t–Statistic	Prob.
C	−2.1894	0.077045	−28.4172	0
TRAN_D (-1)	−0.4922	0.01692	−29.0893	0
AREA	0.004757	0.000163	29.12933	0
EX_D	0.27596	2.806626	0.098324	0.9217
GDP_D	−8.93E−07	2.87E−06	−0.31172	0.7553
ROAD_DUMMY_1	0.089678	0.032791	2.734872	0.0063
SO2*FLOOR_DUMMY_ 1_LOW_0_HIGH	−0.00732	0.00376	−1.94634	0.0518
NO2*FLOOR_DUMMY_ 1_LOW_0_HIGH	−0.00133	0.000783	−1.69653	0.09

(continued)

Table 7.6 (continued)

Variable	Coefficient	Std. error	t–Statistic	Prob.
O3*FLOOR_DUMMY_1_ LOW_0_HIGH	−0.00218	0.000707	−3.07876	0.0021
RSP*FLOOR_DUMMY_1_ LOW_0_HIGH	0.001833	0.000963	1.90333	0.0572
EX_D (-1)	5.493084	2.802137	1.96032	0.0501
GDP_D (-1)	1.76E–07	2.87E–06	0.061384	0.9511
R-squared	0.514025	Mean dependent var		0.001195
Adjusted R-squared	0.510888	S.D. dependent var		0.463427
S.E. of regression	0.324104	Akaike info criterion		0.591467
Sum squared resid	178.9945	Schwarz criterion		0.629563
Log likelihood	−495.479	Hannan-Quinn criterion		0.605563
F-statistic	163.8504	Durbin-Watson stat		1.956388
Prob (F-statistic)	0			

Model 3

Dependent variable: TRAN_D
Method: Least Squares
Sample (adjusted): 3 1822
Included observations: 1716 after adjustments

Table 7.7 Results of Hedonic Pricing model 3

Variable	Coefficient	Std. error	t–Statistic	Prob.
C	−2.19048	0.076671	−28.5699	0
TRAN_D (-1)	−0.49369	0.016842	−29.313	0
AREA	0.004778	0.000163	29.38512	0
EX_D	0.209142	2.793046	0.074879	0.9403
GDP_D	−1.02E–06	2.85E–06	−0.35814	0.7203
ROAD_DUMMY_1	0.089188	0.032632	2.733163	0.0063
FLOOR_DUMMY_1_ LOW_0_HIGH	−0.19321	0.045968	−4.20313	0
SO2*FLOOR_DUMMY_1_ LOW_0_HIGH	−0.00801	0.003745	−2.13976	0.0325
NO2*FLOOR_DUMMY_1_ LOW_0_HIGH	0.001257	0.000993	1.265386	0.2059
O3*FLOOR_DUMMY_1_ LOW_0_HIGH	−0.00122	0.00074	−1.64917	0.0993
RSP*FLOOR_DUMMY_1_ LOW_0_HIGH	0.001545	0.000961	1.607782	0.1081
EX_D (-1)	5.255006	2.789108	1.884117	0.0597

GDP_D (-1)	1.22E–07	2.85E–06	0.042602	0.966

R-squared	0.519015	Mean dependent var	0.001195
Adjusted R-squared	0.515626	S.D. dependent var	0.463427
S.E. of regression	0.322531	Akaike info criterion	0.582312
Sum squared resid	177.1567	Schwarz criterion	0.623583
Log likelihood	−486.624	Hannan–Quinn criterion	0.597583
F-statistic	153.1375	Durbin–Watson stat	1.951517
Prob (F-statistic)	0		

Model 4

Dependent Variable: TRAN_D
Method: Least Squares
Sample (adjusted): 3 1822
Included observations: 1806 after adjustments

Table 7.8 Results of Hedonic Pricing model 4

Variable	Coefficient	Std. error	t–Statistic	Prob.
C	−2.17377	0.077236	−28.1446	0
TRAN_D (-1)	−0.49349	0.016442	−30.015	0
AREA	0.004737	0.000157	30.08828	0
FLOOR_DUMMY_1_ LOW_0_HIGH	−0.16005	0.041685	−3.83959	0.0001
ROAD_DUMMY_1	0.088556	0.031474	2.81367	0.005
SO2*FLOOR_DUMMY_1_ LOW_0_HIGH	−0.00839	0.003789	−2.21351	0.027
O3*FLOOR_DUMMY_1_ LOW_0_HIGH	−0.00201	0.000859	−2.34102	0.0193
RSP*FLOOR_DUMMY_1_ LOW_0_HIGH	0.00323	0.001073	3.01071	0.0026
SO2	0.004022	0.002341	1.717783	0.086
O3	0.000815	0.000495	1.645581	0.1
RSP	−0.00148	0.000639	−2.3091	0.0211

R-squared	0.515407	Mean dependent var	0.001009
Adjusted R-squared	0.512707	S.D. dependent var	0.456213
S.E. of regression	0.318466	Akaike info criterion	0.55547
Sum squared resid	182.0499	Schwarz criterion	0.588963
Log likelihood	−490.589	Hannan-Quinn criterion	0.567831
F-statistic	190.9138	Durbin-Watson stat	1.950827
Prob (F-statistic)	0		

7.6 References

Allen, F. and E. Carletti, (2009) The Global Financial Crisis: Causes and Consequences. *European University Institute.*

Anstine, J. (2003) Property Values in a Low Populated Area When Dual Noxious Facilities are Present. *Growth and Change,* 34, 345–358.

Bourassa, S. C., M. Hoesli and V. S. Peng (2003) Do Housing Submarkets Really Matter? *Journal of Housing Economics,* 12, 12–28.

Bowen, W. M., B. A. Mikelbank and D. M. Prestegaard (2001) Theoretical and Empirical Considerations Regarding Space in Hedonic House Price Estimation. *Growth and Change,* 32, 466–490.

Brassington, D. M. (1999) Which Measures of School Quality Does the Housing Market Value? *Journal of Real Estate Research,* 18, 395–413.

Can, A. (1990) The Measurement of Neighbourhood Dynamics in Urban House Prices. *Economic Geography,* 66, 254–272.

Cebula, R. J. (2009a) The Hedonic Pricing Model Applied to the Housing Market of the City of Savannah and its Savannah Historic Landmark District. *The Review of Regional Studies,* 39, 9–22.

Cebula, R. J. (2009b) The Hedonic Pricing Model Applied to the Housing Market of the City of Savannah and its Savannah Historic Landmark District. *The Review of Regional Studies,* 39, 14.

Centaline Property Agency (2014) *Centadata.*

Chau, K. W., V. S. M. Ma and D. C. W. Ho (2001) The Pricing of "Luckiness" in the Apartment Market. *Journal of Real Estate Literature,* 9, 31–40.

Dale-Johnson, D. (1982) An Alternative Approach to Housing Market Segmentation Using Hedonic Price Data. *Journal of Urban Economics,* 11, 311–332.

David, E. C. (2006) Externality Effects on Residential Property Values: The Examples of Noise Disamenities. *Growth and Change,* 37, 460.

Goodman, A. C. (1978a) Hedonic Prices, Price Indices and Housing Markets. *Journal of Urban Economics,* 5, 471–484.

Goodman, A. C. (1978b) Hedonic Prices, Price Indices and Housing Markets. *Journal of Urban Economics,* 5, 14.

Goodman, A. C. and T. G. Thibodeau (2003) Housing Market Segmentation and Hedonic Prediction Accuracy. *Journal of Housing Economics,* 12, 181–201.

HKEPD – Kwun Tong Monitoring Station (2013) *Kwun Tong Monitoring Station.*

Kong, F., H. Yin and N. Nakagoshi (2007) Using GIS and Landscape Metrics in the Hedonic Price Modeling of the Amenity Value of Urban Green Space: A Case Study in Jinan City, China. *Landscape and Urban Planning,* 79, 240–252.

Kryvobokov, M. (2013) Hedonic Price Model: Defining Neighborhoods with Thiessen Polygons. *International Journal of Housing Markets and Analysis,* 6, 79–97.

Laurice, J. and R. Bhattacharya (2005a) Prediction Performance of a Hedonic Pricing model for Housing. *The Apprasial Journal,* 73, 12.

Laurice, J. and R. Bhattacharya (2005b) Prediction Performance of a Hedonic Pricing model for Housing. *The Appraisal Journal,* 73, 198–209.

Li, R. Y. M. (2009) The Impact of Climate Change on Residential Transactions in Hong Kong. *The Built and Human Environment Review,* 2, 11–22.

Li, R. Y. M. (2012) The Internalization of Environmental Externalities Affecting Dwellings: A Review of Court Cases in Hong Kong. *Economic Affairs,* 32 81–87.

Li, R. Y. M. (2014) *Law, Economics and Finance of the Real Estate Market – A Perspective of Hong Kong and Singapore*. Germany: Springer.

Li, R. Y. M. and Y. L. Li (2011) Judges' View on the Price of Environmental Externalities in the United Kingdom. *US–China Law Review*, 8 994–1007.

Li, R. Y. M., C. Y. Law and T. H. Leung. (2014) "Hong Kong People are No Longer Superstitious? The Pricing of Residential Units' Luckiness Revisit." In *19th Asian Real Estate Society*, Brisbane, Australia.

Limsombunchai, V., C. Gan and M. Lee (2004a) House Price Prediction: Hedonic Price Model vs. Artificial Neural Network. *American Journal of Applied Sciences*, 1, 9.

Limsombunchai, V., C. Gan and M. Lee (2004b) House Price Prediction: Hedonic Price Model vs. Artificial Neural Network. *American Journal of Applied Sciences*, 1, 193–201.

Maclennan, D. and Y. Tu (1996) Economic Perspectives on the Structure of Local Housing Systems. *Housing Studies*, 11, 387–406.

Nelson, J. P. (2004) Meta-Analysis of Airport Noise and Hedonic Property Values: Problems and Prospects. *Journal of Transport Economics and Policy*, 38, 1–28.

Pennington, G., N. Topham and R. Ward (1990) Aircraft Noise and Residential Property Values Adjacent to Manchester International Airport. *Journal of Transport Economics and Policy*, 24, 49.

Rosen, S. (1974) Hedonic Prices and Implicit Markets: Product Differentiation in Pure Competition. *Journal of Political Economy*, 82, 34–55.

Simons, R. A. (2001) The Effects of an Oil Pipeline Rupture on Single-family House Prices. *The Appraisal Journal*, 410–418.

Sirmans, G. S., D. A. Macpherson and E. N. Zietz (2005a) The Composition of Hedonic Pricing. *Journal of Real Estate Literature*, 13, 3–43.

Sirmans, G. S., D. A. Macpherson and E. N. Zietz (2005b) The Composition of Hedonic Pricing Models. *Journal of Real Estate Literature*, 13, 41.

Tekel, A. and L. Akbarishahabi (2013) Determination of Open-green Space's Effect on Around House Prices by Means of Hedonic Price Model; In Example of Ankara/ Botanik Park. *Gazi University Journal of Science*, 26, 347–360.

Witte, A. D., H. J. Sumka and H. Erekson (1979) An Estimation of a Structural Hedonic Price Model of the Housing Market: An Application of Rosen's Theory of Implicit Markets. *Econometrica*, 47, 1151–1173.

8 The impact of the subprime financial crisis on the German and Norwegian real estate markets

L1, the Chow test and the Quantile regression approach

Rita Yi Man Li and Joe Fuk Kin Wong

8.1 Historical background of the subprime financial crisis

The financial crises of recent decades – the global subprime crisis, the Asian financial crisis and the Eurozone debt crisis – led to an economic downturn in some cities and countries. The recent subprime financial crisis during 2007–2009 was considered by many economists to be the worst financial crisis since the first global financial crisis – the Great Depression in the 1930s. The financial turmoil was dubbed by most of the economists as a "subprime" financial crisis as its main cause was the subprime mortgage market in the US. Greedy bankers lent too much money to homebuyers who should not have been able to borrow. The formation of the housing bubble in the US and the blooming of the subprime mortgage market were closely intertwined to breed the recent financial crisis gradually. Moreover, the Federal Reserve lowered its target federal funds rate progressively to 1% in 2003 after the stock market peaked in 2000. The low interest rates motivated investors to look for high yield investments. Collateralized debt obligations, backed by subprime mortgages, offered good investment opportunities to investors. They appeared to be safe investments since borrowers in financial distress could refinance the property to repay the mortgage debt (Li 2012).

Moreover, US housing prices increased significantly from 1998 to 2006. Real housing prices increased by 31.6% from the fourth quarter of 2002 to the fourth quarter of 2006 (Li and Li 2012). According to the S&P/Case Shiller US National Home Price Index, US housing prices took about 18 years to double the price from the national home price index of 46.04 in October 1980 to 92.23 in November 1998. However, it only took another eight years to double to the peak of 184.62 in July 2006.

Sometimes, a rise in housing prices reflects an increase in the demand for housing due to a number of factors which include the sustained lower interest rates (Marshall 2009). For example, the Federal Reserve decreased the federal funds target rates gradually from 6.5% in May 2000 to 1.0% in June 2003. Then the target rates lasted for one more year until June 2004 before the Fed started to increase the rates again gradually to 5.25% in June 2006. During the period of low interest rates, the adjustable-rate mortgages were very attractive to potential

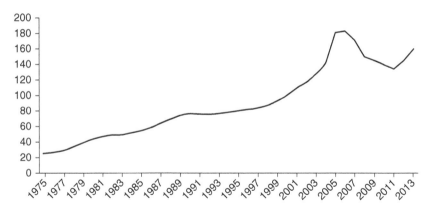

Figure 8.1 S&P/Case Shiller US National Home Price Index, 1975–2014 (S&P/Case
Shiller US National Home Price Index, S&P Dow Jones Indices, 2014)

buyers and the substantial increase in the demand for housing created the housing
bubble afterwards.

According to the 2007 Annual Report of the Federal Reserve Bank of San
Francisco, the subprime mortgage is defined as "a lender-given designation for
loans extended to borrowers with some sort of credit impairment." The subprime
loans and alt-A loans are two main categories of general class of nonprime loans.
Compared with subprime borrowers, alt-A borrowers have higher credit ratings
and are viewed as having lower risk than subprime loans. Figure 8.2 shows the
trend of the distribution of the subprime loans and alt-A loans.

The subprime market began to bloom in the late 1990s. According to the
Annual Report of State of the Nation's Housing 2008, subprime and alt-A originations still only accounted for 7.2% and 2.5% of total mortgage originations in
2008 respectively. By 2006, the subprime and alt-A originations had reached the
peak of accounting for 20.2 and 13.4% respectively. There are cost and benefit
of getting subprime loans. The cost is the substantially higher interest rates compared with that of prime loans. In view of this, the subprime mortgage market can
be considered as using a risk-based pricing method to allocate loans instead of
nonprime credit rationing by setting strict minimum lending standards. The benefits are the increased numbers of hard-to-qualify borrowers who are subprime
loan lenders and thus the opportunity for them to create wealth (Chomsisengphet
and Pennington-Cross 2006).

The "originate and distribute model" and the securitization of subprime mortgage loans is largely the culprit of the subprime financial crisis. The subprime
lending is a lucrative business for the originators as they have the incentives to
sell as many mortgages as possible and then sell them off to financial institutions
such as Fannie and Freddie, without bothering if the borrowers can repay the loan
or not. The subprime mortgages loans are then securitized into so-called high

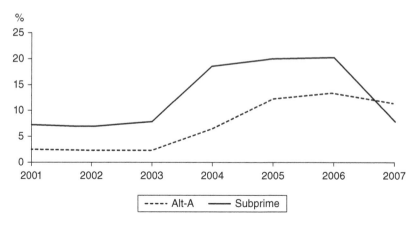

Figure 8.2 Mortgage Originations by Product: 2001–2007 (Joint Center for Housing Studies of Harvard University 2008)

quality (AAA) mortgage-backed securities (MBS) for investors who desire to find financial instruments with low risk and high returns (Allen and Carletti 2009).

However, the lucrative business posed a serious risk of moral hazard. As economist Joseph Stiglitz (2008) summarized:

> [s]ecuritization was based on the premise that a fool was born every minute . . . Globalization meant that there was a global landscape on which they could search for those fools – and they found them everywhere. Mortgage originators didn't have to ask, is this a good loan, but only, is this a mortgage I can somehow pass on to others.

The subprime mortgage market bloom and then the US housing bubble burst had contagion effect on different sectors of the US economy as well as on many countries around the world. As the number of foreclosures increased dramatically, credit rating agencies started to downgrade the ratings of many asset-backed financial instruments. As the large credit rating companies inflate the ratings of MBS, collateralized debt obligations (CDOs), and other related derivatives, they were blamed as one of the major causes of the crisis. As reported in BusinessWeek on September 30, 2007: " . . . in a single day, S&P slashed its ratings on two sets of AAA bonds backed by residential mortgage securities to CCC+ and CCC, instantly changing their status from top quality to pure junk." The collapse of the subprime mortgage market is a prelude to the global financial crisis during 2007–09. The crisis began as housing prices in the US started to fall from the national home price index of 184.62 in July 2006 to the trough of 134.05 in February 2012, lasting for about six years with a decrease of 27.4%, which is more than twice that recorded during the Great Depression.

8.2 The impact of the financial crisis on the housing market

Having studied the historical record, Reinhart and Rogoff (2008a) used the data of banking and financial crises around the world as early as 1800. They explained the concept of "this-time-is-different" syndrome as "rooted in the firmly held belief that financial crises are things that happen to other people in other countries at other times; crises do not happen to us, here and now. We are doing things better, we are smarter, we have learned from past mistakes. The old rules of valuation no longer apply" (Reinhart and Rogoff 2009b).

Claessens and Kose (2013) identified four types of financial crises, namely currency crises, sudden stops crises, external and domestic crises, and banking crises. They summarized four general features about the causes of financial crises: asset price increases substantially, credit booms, build-up of systemic risk, and imprudent regulation and supervision. With regards to the recent global crisis, in particular, they pointed out four new causes:

1   the widespread use of derivative instruments;
2   the strong contagion effect of financial crisis;
3   the high level of leverage of financial institutions;
4   the excessive leverage of the household sector.

Reinhart and Rogoff (2008b) studied asset price bubbles and banking crises. By analyzing the magnitude and duration of the downturn in housing prices that has historically accompanied major banking crises in both advanced and emerging economies, they concluded that the duration of the cycle in real housing prices was about four to six years and the magnitudes of the declines in real housing prices around banking crises from peak to trough of the emerging economies is not significantly different from that of the advanced economies, which strongly supported their contention that "banking crises are an equal opportunity menace." In their study, they pointed out that the average of the cumulative decline in real housing prices from peak to trough was 35.5% and Hong Kong had 58.9% decline in real housing prices during Asian financial crisis of 1997–1998. Table 8.1 illustrates the magnitude and duration of the downturn in housing prices of some countries with a drop of more than 40% (Reinhart and Rogoff 2008b).

To assess the relationship between credit growth and real estate prices in Asian economies during the Asian financial crisis, both panel regressions and individual country Value-at-Risk (VaR) measures were adopted by Collyns and Senhadji (2003). They suggested that property prices were strongly procyclical and bank lending which significantly contributed to the increase of property price prior to the crisis. Moreover, considering the response of property prices to credit, it was significantly stronger before the crisis and asymmetric during periods of rising and declining property prices. Fung and Forrest (2002) explored the institutional structure of the housing market and its effect on the housing prices in Hong Kong during the Asian Financial Crisis. They studied the role of the

Table 8.1 Real housing price cycles and banking crises (Reinhart and Rogoff 2008b)

Country	Crisis date	Peak	Trough	Duration of downturn	Magnitude of decline (in %)
Advanced economies					
Finland	1991	Quarter 2 in 1989	Quarter 4 in 1995	6 years	−50.4
Japan	1992	Quarter 1 in 1991	Ongoing	Ongoing	−40.2
Norway	1987	Quarter 2 in 1987	Quarter 1 in 1993	5 years	−41.5
Asian financial crisis					
Hong Kong	1997	Quarter 2 in 1997	Quarter 2 in 2003	6 years	−58.
Indonesia	1997	Quarter 1 in 1994	Quarter 1 in 1999	5 years	−49.9
Philippines	1997	Quarter 1 in 1997	Quarter 3 in 2004	7 years	−53.0
Other emerging economies					
Colombia	1998	Quarter 1 in 1997	Quarter 2 in 2003	6 years	−51.2

giant-developer-dominated structure and the interaction between the government and the developers.They concluded that the immense resources possessed by the giant developers limit the effects and scope of the government's continual intervention.

8.3 The impact of the subprime financial crisis on the housing market: a global perspective

Financial crises shared many similarities such as the dramatic decrease of asset prices, the huge accumulation of debt, the downturn of economic growth, the increase of current account deficits (Reinhart and Rogoff 2008a) and significant of money lose in the financial market (Li 2014). Nevertheless, the impact is uneven (Li 2011). The downturn of economic growth is shown in Table 8.2. It indicates the decline in economic growth during the subprime financial crisis for country groups and some selected economies in that country group. We calculated the growth decline as the absolute value of the difference in real GDP growth rates between 2007 and 2009.

Many countries' economic growth decreased during this crisis, ranging from the smallest decline of 3.1% in Middle East and North Africa to the largest decline of 9% in Central and Eastern Europe. In addition, some countries recorded substantial decline in economic growth, including the United Arab Emirates, Turkey, Hong Kong and Germany, whereas some countries faced a small decline in economic growth – for example, India, Norway, China and the United States. Some may argue that Asian economies are decoupled from the crisis which erupted in the United States. However, the impact on economic growth revealed the strong contagion effect of the financial crisis due to the extent of trade and financial market exposure of the Asian economies, especially the Asian newly industrialized economies (Goldstein and Xie 2009).

Table 8.2 Economic growth slowdown in 2007–2009 (IMF WEO database 2014)

Country group name	2007	2008	2009	2007–2009 change
World	5.4	2.8	−0.6	−6.0
Advanced economies	2.8	0.1	−3.5	−6.3
Norway	2.7	0	−1.6	−4.3
Euro area	3.0	0.4	−4.4	−7.4
Germany	3.4	0.8	−5.1	−8.5
Major advanced economies (G7)	2.3	−0.4	−3.8	−6.1
United States	1.9	−0.3	−3.1	−5.0
Newly industrialized Asian economies	5.9	1.8	−0.7	−6.6
Hong Kong	6.4	2.3	−2.6	−9.0
Emerging market and developing economies	8.7	6.1	2.7	−6.0
China	14.2	9.6	9.2	−5.0
Central and eastern Europe	5.4	3.12	−3.6	−9.0
Turkey	4.7	0.7	−4.8	−9.5
Developing Asia	11.4	7.9	7.0	−4.4
India	10.0	6.9	5.9	−4.1
ASEAN-5	6.3	4.8	1.7	−4.6
Malaysia	6.3	4.8	−1.5	−7.8
Middle East and North Africa	5.7	4.5	2.6	−3.1
United Arab Emirates	6.6	5.3	−4.8	−11.4

Khor and Kee (2008) compared the recent subprime financial crisis and the Asian financial crisis during 1997–1998. They believed that both crises have similar causes in the sense that they were triggered by investor panic, liquidity run and increasing insolvency in the banking system. Both crises had excessive liquidity and imprudent credit booms, leading to the dramatic increase in housing prices and the burst of the housing bubbles.

Al-Malkawi and Pillai (2013) compared the performance of pre- (2006–2007) and post-crisis (2008–2009) in real estate and construction sector in the United Arab Emirates. Fourteen financial ratios of six companies in the real estate and construction sector have been analyzed to study the liquidity, profitability, leverage and activity position of these companies during the pre- and post-crisis period. By using the Wilcoxon matched-pairs signed rank test, they concluded that there has been a decline in the overall liquidity, profitability leverage and activity positions of the companies after the financial crisis, leading to a fall in the real estate prices in the UAE after the global financial crisis. The major reasons were: oversupply of property before the crisis and then the subsequent suspensions of the projects, huge fluctuations in rental income and negative cash flow. However, there was only mild impact on Turkey due to its small economy, the lack of securitization product markets, and the inefficient housing credit market (Coşkun 2011).

The impact of the subprime financial market on China's housing market started in November 2007. Figure 8.3 shows the trend of the Chinese Housing Prosperity Index (CHPI). The index fell from 106.59 in November 2007 to 94.74 in March 2009 which last for 15 months with a mild drop of 11.1% (China Real Estate Information 2009).

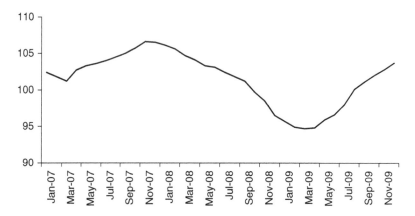

Figure 8.3 Chinese Housing Prosperity Index (CHPI) from 2007 to 2009 (China Real Estate Information, 2009)

China also faced two conditions similar to the reasons for the subprime crisis in the US, which included aggressive interest rate adjustments and over-lending activities of the banking and non-bank mortgage originators. However, China did not have the complex asset securitization, which explains why it was not so seriously affected by the crisis (Yao *et al.* 2010).

8.4 Data

We collected the yearly data of Gross Domestic Product (GDP), export, import, terms of trade (calculated by export value divided by import value), labor force, interest rate on deposit, construction new orders (residential) data and quarterly data of housing price index (HPI) from Statistics Norway (2014).

In Germany, we used the monthly data of unemployment rate, foreign investment, terms of trade, production index, new orders of production, quarterly data of GDP and housing price index to construct the econometric models.

8.5 Research method

As there were a number of economic and social variables with lower frequency data than the others, we first adopted Cubic Spline Interpolation to find the points between given data intervals. The idea of the interpolation starts from the points $[a_i b_i]$ [where $i = 0,1, \ldots$, n for the function b = f(a), n + 1 points with n intervals between them. The cubic spline interpolation is a piecewise continuous curve which pass through each of the values in the table (Drakos and Moore, 2002). According to Drakos and Moore (2002), there is a separate cubic polynomial for each interval, each with its own coefficients:

$$S_i(a) = x_i(a-a_i)^3 + y_i(a-a_i)^2 + z_i(a-a_i)^2 + w \text{ where } a \in [a_i, a_{i+1}]$$

These polynomial segments are then denoted as the spline S(a). Since there are n intervals with four coefficients for each we require a total of 4n parameters to define the spline S(a). We need to find 4n independent conditions to fix them. We have two conditions for each interval such that the cubic polynomial matches the values of the table at both ends of the interval:

$$S_i\left(a_i\right) = b_i \text{ and } S_i\left(a_{i+1}\right) = b_{i+1}$$

As these conditions result in a piecewise continuous function, 2n more conditions are needed. To smoothen the interpolation, the first and second derivatives have to be continuous:

$$S''_{i-1}\left(a_i\right) = S'_i\left(a_i\right) \text{ and } S''_{i-1}\left(a_i\right) = S''_i\left(a_i\right)$$

where the conditions i = 1, 2, . . . , n−1 applied with 2(n−1) constraints, two more conditions have to be fulfilled to fix the spline:

$$S'_0\left(a_0\right) = f'\left(a_0\right) \text{ and } S''_{n-1}\left(a_n\right) = f'\left(a_n\right)$$

$$S''_0\left(a_0\right) = 0 \text{ and } S''_{n-1}\left(a_n\right) = 0$$

The following tables summarize the data after adopting the cubic spline interpolation. Quartile regression Chow test and L1 analysis was then used to analyze the data.

After that, the data was then dealt with Least Absolute Deviation (L1) regression/ Quantiles regression. According to Kim *et al.* (2010), the multivariate linear regression is presented as:

$$\tilde{y} = a_0 + \sum_{i=0}^{n} a_i x_i$$

In L1, the sum of absolute errors is calculated by the absolute values of the difference between the original and predicted values as:

$$SAE = \sum_{i=0}^{n} |y_i - \tilde{y}_i| = \sum_{i=0}^{n} |y_i - a_0 - \sum_{i=0}^{n} a_i x_{ij}|$$

where n denotes the training period and y_i represents the original values, others are the same as the multivariate linear regression.

Quantiles are related to the operations of ordering the sample observations, such that we can define the quantiles via a simple alternative expedient as an optimization problem. We define the sample mean as the solution to minimize a sum of squared residuals, and the median as the solution for the minimum value of the sum of absolute residuals. The symmetry of the piecewise linear absolute value function implies that the minimization of the sum of absolute residuals must equate the number of positive and negative residuals such that there are the

Table 8.3 Summary of Norway data

Variable	Mean	Median	Minimum	Maximum	Std. dev.	C.V.	Skewness	Ex. kurtosis
HPI	93.3692	83.9000	32.5000	174.000	44.1396	0.472743	0.279677	-1.20847
GDP	2.44593e+011	1.98000e+011	-1.17000e+012	1.10000e+012	2.55592e+011	1.04497	-2.06418	12.1315
Exports	9.89519e+010	8.03856e+010	-6.13000e+010	5.39000e+011	1.23615e+011	1.24925	-2.36152	14.7623
Imports	6.95272e+010	5.46647e+010	-4.10000e+010	3.55000e+011	8.21223e+010	1.18115	-2.54677	15.6338
CPI (Year2010 = 100)	85.1093	85.8366	36.5226	117.306	13.4092	0.157553	-0.375126	0.774367
Labor force in total (1000 persons)	2397.19	2373.00	2104.00	2721.00	168.929	0.0704697	0.105897	-0.877019
Interest rates on deposits	3.70451	3.36000	1.25000	9.05000	1.85392	0.500450	0.883906	0.370025
Construction New Orders (Residential)	72.6923	67.0000	15.0000	129.000	34.8009	0.478743	0.0489557	-1.33481
TOT	1.39949	1.44082	0.989282	2.78079	0.251713	0.179861	2.67246	12.7688

Table 8.4 Summary of Germany data

Variable	Mean	Median	Minimum	Maximum	Std. dev.	C.V.	Skewness	Ex. kurtosis
GDP (billion dollars)	648.428	641.770	577.300	718.470	37.1707	0.0573244	0.0699272	−0.918950
Unemployment Rate	8.01354	7.60000	6.40000	12.2000	1.37195	0.171203	1.24141	1.09668
Foreign Investment (Unit in million)	14703.4	10967.0	−79737.0	94072.0	38714.7	2.63305	−0.377205	−0.0894295
Terms of trade	1.21160	1.20909	1.12361	1.29231	0.0404396	0.0333769	0.0352655	−0.640507
Residential Price Index (2010 = 100)	101.756	99.7500	95.4000	111.000	4.60052	0.0452112	0.816918	−0.643189
Production index	103.008	104.150	80.8000	117.500	8.84805	0.0858964	−0.555382	−0.142096
New Orders	102.389	104.150	72.9000	123.000	10.9479	0.106925	−0.760387	0.473742
Log residential price index 2010_100	4.62159	4.60267	4.55808	4.70953	0.0444567	0.00961936	0.770805	−0.706156

same number of observations above and below the median. After that, minimizing a sum of asymmetrically weighted absolute residuals – simply giving differing weights to positive and negative residuals – would yield the quantiles (Koenker and Hallock 2001):

$$\min_{\partial \in R} \sum \rho_\gamma \left(y_i - \partial \right)$$

where the function ρ_y is the tilted absolute value function which yields γth sample quantile as the solution. Least square regression offers a model with sample $y_1, y_2, \ldots y_n\}$ to solve the equation:

$$\min_{\partial \in R} \sum_{i=1}^{n} \left(y_i - \mu \right)^2$$

we obtain the sample mean, an estimate of the unconditional population mean (Ey).
If we now replace the scalar m by a parametric function μ (x, b) and solve:

$$\min_{b \in R} \sum_{i=1}^{n} \left(y_i - \mu\left(x, b\right) \right)^2$$

We then obtain the conditional expectation E(y|x).
In Chow test, the regression $y_t = a_0 + B_t X_t + e_t$ is split into two halves if structural break exists, such that:

$$y_{t1} = a_{t1} + B_{t1} X_{t1} + e_{t1}$$

$$y_{t2} = a_{t2} + B_{t2} X_{t2} + e_{t2}$$

The first set of question applies before the structural break and the second equation denotes the model after the structural break. If all the parameters are the same $B_{t1} = B_{t2}$ and $a_{t1} = a_{t2}$. In essence, Chow test tests whether one equation or two equations fit the model best (Li 2012).

8.6 Results

The following tables show the results of Norway and Germany's housing price models. In Norway, results of L1 show that the impact of the subprime financial crisis from December 2007 to June 2009 as a whole was not significant. Nevertheless, the results of Chow Test suggest that a structural break did exist in quarter four of 2008. The four models show that construction orders (six and seven months ahead of housing price) were of significant results in L1 and Chow test.

Table 8.5 L1 and Chow tests results for housing prices in Norway

Dependent variable: housing price index

Model number	1	2	3	4
Modeling method	L1	L1	Chow Test	Chow Test
Constant	−385.779***	−375.885***	−123.959***	−126.985***
GDP_current US	−9.7391e−012	5.9417e−011	−1.10815e−010	−3.78802e−011**
Consumer price index (2010=100)	0.362603	0.0953822	1.03494**	1.07130**
Labor force in total (1000 persons)	0.179087***	0.180774***	0.0401616	0.0390464*
Interest rates on deposits	−1.70875**	−1.30005	−1.47765**	−1.17375**
Construction New Orders (lead of six months)	0.412328***			0.624323***
Construction New Orders (lead of seven months)		0.433682***	0.654359***	
Terms of trade	−1.14071			−0.181303
Subprime dummy	−7.93753	−7.24166		
Exports of goods and services		−1.23937e−010	1.46585e−010	
Split dummy			−540.558**	−499.194***
sd_GDP~			1.71778e−010	−8.32108e−011***
sd_Consumer price index~			−1.00657	2.42390***
sd_Labor force ~			0.265922***	0.118148**
sd_Interest rates~			7.30682 ***	2.86327
sd_Construction~			−0.719692***	−0.448713***
sd_Exports of good~			−2.81438e−010	
sd_Terms of trade				32.7644***
Chow test for structural break at observation			2008 Q4	2008 Q4
Chow Test p-value			0.0000	0.0000

Note: Models 3 and 4 are selected out of the Chow test results of quarters one to six of 2008 (Models 5 and 6 are in singular matrix).
*** represents significant at 99% level, ** refers to significant at 95% level and * denotes significant at 90% level.

Table 8.6 Chow test and Quantile regression for housing prices in Germany

Dependent variable: log housing price index

Model	1	2	3	4
	Chow test	*Chow test*	*Chow test*	*Quantile regression*
Constant	175.742***	5.27877***	5.24302***	4.84840***
GDP	−0.0819777**	−0.000699365*	−0.000656666*	−0.000369789***
Unemployment rate	−1.17423*	−0.00910633	−0.00794623	−0.00411541***
Foreign Investment	8.66631e−06	1.01743e−07	9.45739e−08	2.26505e−08
Terms of Trade	−10.4372	−0.130766	−0.130272	0.0142049
New Orders	−0.118303**	−0.00119191**	−0.00139002***	−4.05158e−05
Production Index	0.0920752	0.00107991*	0.00125818**	−2.85183e−05***
Subprime dummy				−0.00862815***
Split dummy	−61.4615	−0.545321	−0.502759	
sd_GDP	0.0666459	0.000573085	0.000531172	
sd_Unemployment rate	−0.716780	−0.00980757	−0.0117037	
sd_Foreign Investment~	−3.15702e−07	−0.00980757	−1.13270e−08	
sd_Terms of trade~	23.7458*	0.265464**	0.266101	
sd_New orders	0.215608***	0.00207031***	0.00224258***	
sd_Production Index~	−0.254434 ***	−0.00266239***	−0.00284646***	
Structural break	Quarter 4 in 2008	Quarter 3 in 2008	Quarter 2 in 2008	
P-value	0.0002	0.0002	0.000	
TAU				0.5

Note: The model in 2008 quarter 5's p-value was 0.09% which was the highest and was not included in the table.
*** represents significant at 99% level, ** refers to significant at 95% level and * denotes significant at 90% level.

For example, labor force changed from insignificant to significant at 99% level in model 3. Terms of trade changed from insignificant to significant at 99% level in model 4. In Germany, the results of the Chow test shows that GDP changed from significant to insignificant. The quartile regression's subprime dummy was negative and significant, indicating that the averse impact of the financial crisis on housing prices.

8.7 Conclusion

The subprime financial crisis had a negative impact on many countries. There was a rise in unemployment in the US, a reduction in demand for the products made in the Eurozone area which led to a worsened economy in many European countries. The negative externalities on the European countries did not rest only on poor exports but also on the housing markets. Germany was no exception.

The research results show that housing prices dropped during the subprime financial crisis as confirmed by the subprime dummy in Quantile regression. The existence of a structural break further confirmed the results. On the other hand, although the general overall impact on Norway's housing price was insignificant during the whole period of the subprime financial crisis, the results of the Chow test suggested that a structural break exists nevertheless.

8.8 References

Allen, F. and E. Carletti (2009) The Global Financial Crisis: Causes and Consequences. *European University Institute*, 30 June 2009.

Al-Malkawi, H. N. and R. Pillai (2013) The Impact of Financial Crisis on UAE Real Estate and Construction Sector: Analysis and Implications. *Humanomics,* Vol. 29, No.2, 115–135.

BusinessWeek (2007) Anatomy of A Ratings Downgrade. *BusinessWeek,* 30 September 2007.

China Real Estate Information (2009) China Real Estate Information http://www.reales-tate.cei.gov.cn/tj/index.aspx

Chomsisengphet, S. and A. Pennington-Cross (2006) The Evolution of the Subprime Mortgage Market. *Federal Reserve Bank of St. Louis Review*, January/February 2006. 88(1), 31–56.

Claessens, S. and M. A. Kose (2013) Financial Crises: Explanations, Types, and Implications. *IMF Working Paper*, WP/13/28.

Collyns, C. and A. Senhadji (2003) "Lending Booms, Real Estate Bubbles, and the Asian Crisis." In Hunter, W. C., Kaufman, G. G. and Pomerleano, M. (Eds). *Asset Price Bubbles: The Implications for Monetary, Regulatory, and International Policies.* The MIT Press.

Coşkun, Y. (2011) The Global Financial Crisis and the Turkish Housing Market: Is There a Success Story? *Housing Finance International*, Spring 2011, 6–15.

Drakos, N. and R. Moore (2002) Cubic Spline Interpolation. http://www.physics. utah/~detar/phys6720/handouts/cubic_spline/node1.html

Federal Reserve Bank of San Francisco (2007) The Subprime Mortgage Market: National and Twelfth District Developments. *Federal Reserve Bank of San Francisco*, 2007 Annual Report.

Fung, K. K. and R. Forrest (2002) Institutional Mediation, The Hong Kong Residential Housing Market and the Asian Financial Crisis. *Housing Studies,* Vol. 17, No. 2, 189–207.

Goldstein, M. and D. Xie (2009) The Impact of the Financial Crisis on Emerging Asia. *Peterson Institute for International Economics*, Working Paper 09–11.

Joint Center for Housing Studies of Harvard University (2008). State of the Nation's Housing 2008 http://www.jchs.harvard.edu/research/publications/state-nations-housing-2008

Khor, Hoe Ee and Rui Xiong Kee (2008) Asia: A Perspective on the Subprime Crisis. *Finance and Development*, June 2008.

Kim H. S., C. H. Ho, P. S. Chub, and J. H. Kim (2010) Seasonal Prediction of Summer Time Tropical Cyclone Activity over the East China Sea using the Least Absolute Deviation Regression and the Poisson Regression. *International Journal of Climatology*, Vol. 30, No. 2, 210–219.

Koenker, R. and K. F. Hallock (2001) Quantile Regression. *Journal of Economic Perspectives,*Vol. 15, No. 4, 143–156.

Joint Center of Housing Studies of Harvard University (2008). The State of the Nation's Housing 2008. *Joint Center of Housing Studies of Harvard University*, Working Papers and Reports.

Li, R. Y. M. (2011) "A statistical review on the uneven impact of subprime financial crisis on global property market," In *International Symposium on Accounting and Finance 2011,* Macau, November 21–24, 2011

Li, R. Y. M. (2012) "Chow Test Analysis on Structural Change in New Zealand Housing Price During Global Subprime Financial Crisis." *18th Annual Pacific Rim Real Estate Society Conference,* Adelaide, Australia, 15–18 January 2012

Li, R. Y. M. (2014) *Law, Economics and Finance of the Real Estate*. Springer.

Li, R. Y. M. and J. Li (2012) "The Impact of Subprime Financial Crisis on Canada and United States Housing Market and Economy." In *International Conference on Business, Management and Governance Conference,* Hong Kong, 29–30 December 2012

Marshall, J. (2009) The Financial Crisis in the US: Key Events, Causes and Responses. *House of Commons Library*, Research Paper 09/34, 22 April, 2009.

Reinhart, C. M. and K. S. Rogoff (2008a) Is the 2007 US Sub-prime Financial Crisis So Different? An International Historical Comparison. *American Economic Review: Papers and Proceedings 2008,* 98:2, 339–344.

Reinhart, C. M. and K. S. Rogoff (2008b) Banking Crises: An Equal Opportunity Menace. *National Bureau of Economic Research*, Working Paper 14587, December, 2008.

Reinhart, C. M. and K. S. Rogoff (2009a) International Aspects of Financial-Market Imperfections: The Aftermath of Financial Crises. *American Economic Review: Papers and Proceedings 2009,* 99:2, 466–472.

Reinhart, C. M. and K. S. Rogoff (2009b) *This Time is Different: Eight Centuries of Financial Folly*. Princeton University Press.

SandP/Case Shiller US National Home Price Index (2014) *SandP Dow Jones Indices.* http://us.spindices.com/indices/real-estate/sp-case-shiller-us-national-home-price-index

Stiglitz, J. (2008) Testimony of Joseph Stiglitz, Hearing on The Future of Financial Services Regulation. *House Committee on Financial Services*, 21 October, 2008.

Yao, S., D. Luo and S. Morgan (2010) Impact of the US Credit Crunch and Housing Market Crisis on China. *Journal of Contemporary China*, 19(64), March, 401–441.

9 Housing prices and external shocks' impact on South Africa and the Czech Republic's housing prices

A Vector Error Correction model and impulse response functions approach

Rita Yi Man Li, Joe Cho Yiu Ng, Kwong Wing Chau and Tat Ho Leung

9.1 Introduction

Partly because of the expansionary monetary policies in many countries in recent decades, housing prices in many mega cities are very high and have been on a rising trend for years (although there have been times when prices have fallen). In many countries housing is considered to be the most important investment category in households' investment portfolios (Hochguerte 2001, Li 2015). In fact, many households' wealth are held in the form of housing (Hochguerte 2001) instead of the traditional gold and silver.

As there is an indispensable relationship between the built environment and our economy, previous research has thrown a light on this area. For example, Malpezzi (1999) studied the relationship between the real estate market and macroeconomy. Becker and Morrison (1999) discussed the patterns of urbanization in developing countries, studied the models of third world city growth via partial equilibrium models, economy-wide computable general equilibrium models, household migration modeling as well as general equilibrium models. Kahn (1931) suggested that an increase in employment is usually connected with an increase in investment. For example, a man who finds a new job devotes a large proportion of his increase in income to home-produced goods.

It is often thought that durable assets play a dual role – as factors of production and important collateral for loans (Kiyotaki and Moore 1997). Housing is no exception. Nevertheless, standard macroeconomics textbooks either treat housing as one of many consumption goods or simply neglect it. The so-called mainstream macroeconomics tends to ignore the housing market. More recently, however, there is a small yet growing research effort which bridges the gap between them (Leung 2004). For example, various studies had been conducted to study the impact of financial crises on global property markets (Li 2011, Li and Chan 2013, Li 2012, Li and Wong 2013) and the relationship between property prices and business cycles (Li 2014, Li and Li 2012).

The Czech Republic and South Africa are both developing countries, yet the former has 2.3 times more money than the latter. We hypothesize that the transmission mechanism is different and we would like to compare and contrast the housing price transmission mechanism in richer and less rich developing countries (Royal Oak Interactive 2015) due to the fact that they have different economic development backgrounds. We attempt to find out whether the wealth effect (housing prices cause consumption) and the balance sheet effect (housing prices cause investment) exist in the housing markets of the Czech Republic and South Africa. This can be tested by applying a Granger causality test and impulse response functions based on VECMs. The second task is to imitate upward interest rates by using impulse response functions in order to know how housing prices react in each country.

9.2 A general overview of the Czech Republic's housing market

As early as 1918, the state supported new housing construction in the Czech Republic. It was accompanied with measures which aimed at tenant protection and housing construction increase. After 1948, pre-war housing cooperatives were "nationalized" by merging their property into new large housing cooperatives and most of the private apartment buildings were expropriated by the state. The state housing policy in this period considered housing as an important good in life and increases in housing costs should not lead to an increase in householders' burdens. This necessarily led to a rise in state subsidies for housing construction, management, operation and maintenance of the existing state housing. State dwellings' rent was fixed at the level of 1964 prices till 1990. Thus, no extra funding is available even for the most basic maintenance activities. Moreover, the whole housing system was characterized by under-maintenance, unskilled management, bureaucracy, huge inefficiencies, low-quality construction, free-riding and encouraged the existence of illegal practices and corruption (Lux 2000). Similar to mainland China, the huge expenditure on housing sector became one of the major driving force in the Czech Republic's privatization process (Li and Chau 2011, Lux 2000).

Furthermore, wider economic liberalization and political changes after the collapse of the socialist regime motivated the whole process. This housing reform led to the rapid introduction of a legislative framework to support the establishment of new private construction companies, privatization of construction firms, and the liberalization of housing construction material prices. The termination of state financed housing construction led to a sharp decrease in rental housing construction after 1991. Yet, the deregulation process in housing market almost stopped in 1999 when the government decided to increase regulated rents by the rate of inflation only. Regulated rents were then frozen at 2002 levels from 2002–2006. This heightened tensions between landlords and tenants and the landlords started to bring charges against the Czech Republic before the Court in Strasbourg (Lux 2000).

9.3 A general overview of the housing market in South Africa

In 1994, South Africa's first elected government inherited an estimated housing backlog of 1.2–2.5 million units with a rapid expansion of informal squatter settlements. The total demand in housing increased from 2.5 million to about three million households from 1994 to 1997. Approximately 1.4 million households live in informal housing conditions and squatter camps. The inherited housing backlog was largely a result of the history of apartheid and separate development, high levels of unemployment and poverty, together with urbanization. The South African housing finance system also suffered from home loans defaults (the default rate was about 15%, estimated to be worth R10 billion in 1994). Soon after 1994, the government initiated a multi-pronged approach to induce mortgage lenders to provide housing loans to the low income market (Pillaya and Naude 2006). In the rental market, the Rental Housing Act [No. 50 of 1999] repealed the previous rent control (Republic of South Africa Government 1999). Housing rent in South Africa is freely determined by the market.

9.4 The relationship between the global financial environment and the housing market

After experiencing an unprecedented economic catastrophe during the global financial crisis in 2008, the global financial environment reached a historically low level of interest rates. Although the global economy is currently improving (in December 2015), another global risk is an increasing possibility. The Global Risk Report of the World Economic forum (World Economic Forum 2010, World Economic Forum 2011) highlighted that an asset price collapse was very likely to occur in the coming years. The low interest rate environment of the global economy cannot be prolonged and there is a strong upward pressure on global interest rates due to the expansion of debt. After the 2008 crisis, central banks in different countries stimulated economic growth by lowering interest rates to a historically low level. This low interest rate policy created the risk of financial distortions, serious inflation and asset price bubbles in emerging markets. To avoid these situations, the global interest rate must rise (Gambacorta and Mistrulli 2011, World Economic Forum 2010, World Economic Forum 2011). Housing price volatility may in turn affect the real economy through the wealth effect and the balance sheet effect.

9.4.1 The impact of the wealth effect on real estate prices

In the US and the UK, changes in the macroeconomy have often been associated with property price changes. Previous studies showed that housing wealth effects exist in many countries. In the UK, increase in housing wealth played a significant role in the consumer boom during the 1980s. In Singapore, where public housing dominates, changes in private housing prices have no significant effect on aggregate consumption. In sharp contrast, however, public housing wealth effects are more persistent and larger (Edelsteina and Lum 2004).

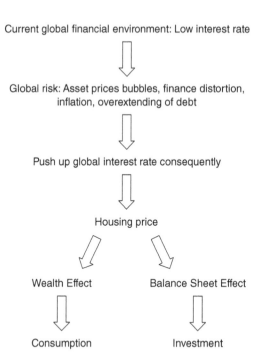

Figure 9.1 The role of housing prices in monetary transmission mechanism during an increase in global interest rate (author's figure)

9.4.2 The balance sheet effect, housing prices and credit market

The balance sheet effect refers to a positive relationship between housing prices and the credit ability of financial institutions (Mishkin 2010). The idea that household debts may worsen economic recession dated back to Fisher's research in 1933. Fisher (1933) explained the unusual severity of the Great Depression was caused by the rising real burden of debt after a general fall in prices (Movshuk 2010). Chen and Patel (1998) examined the causality between housing prices, income, interest rates, stock price index, construction costs, and housing completions in Taipei. Their results showed that there was a feedback mechanism between housing prices and the stock price index. Nguyen and Wang (2010) found that GDP growth affected housing returns but not vice versa. Hence, whether there is a wealth effect and a balance sheet effect in housing prices is still questionable.

9.5 Data description

Among the various macroeconomic factors which drive housing prices, we have included the data of consumer spending and investment expenditure in our model. These two factors have important implications on the housing markets under the

Table 9.1 Data description

Country	Variables	Description	Period	Source
South Africa	hpi_sa	Housing price index	2002 Quarter1 to 2010 Quarter 2	FNB PropertyBarometer Statistic South Africa
	cs_sa	Consumer spending		Statistic South Africa
	inv_sa	Investment expenditure		Statistic South Africa
	int_sa	Prime rate		Statistic South Africa
	sp_sa	JSE All Share Index		Johannesburg Stock Exchange
Czech Republic	hpi_c	Housing price index	1999 Quarter1 to 2009 Quarter 4	CZSO, CNB calculation
	cs_c	Consumer spending		Czech Republic National Bank
	inv_c	Investment expenditure		Czech Republic National Bank
	int_c	Prime rate		Czech Republic National Bank
	sp_c	PX-50 Index		Prague Stock Exchange

lens of Keynesian economics: expenditure is not only an out of pocket activity. The money spent by an individual becomes the income earned by others. Hence, the small sum of monetary expenditure has a multiplier effect.

Furthermore, the prime rate provides useful insights into the costs of borrowing. Households are more willing to borrow when the costs of borrowing are low and vice versa. It often seems that there is a strong linkage between the prime rate (or interest rate) of housing prices as prices are often so high that we do not have sufficient funds to buy without borrowing. Nevertheless, different individuals have different interest rates due to differences in earnings, etc. So, the prime rate should be a good proxy for the interest rate among all homebuyers.

Many of the previous studies include stock prices in their analysis such as Chan and Patel (1998). It is because stock market fluctuations lead to the wealth effect and Tobin's Q effect,[1] which affects consumption and investment. In order to find out the net wealth effect and balance sheet effect of the housing markets, stock prices in South Africa and the Czech Republic will be included in our analysis. The following table shows the description of the data used in this study. Seasonal adjustment is made using Census X12. All the data have been converted to real term and logarithmic form.

---

1 Tobin's Q was introduced by Tobin (1969). It is a ratio between the market value and replacement value of an asset, usually stock prices. When Tobin's Q is greater than one, the market value (stock price) of a company is greater than its assets. The company can issue shares to obtain funds for investing in capital. Thus the higher the stock price, the more the investment.

9.6 Research method and results

The testing procedure includes the following steps. First, a unit root test is conducted on each variable. If all the variables are integrated at order one, I (1), then a cointegration test will be employed. If a cointegrating relationship exists among the variables, then the Granger causality test and impulse response functions analysis can be conducted based on VECM. A cointegration test has been applied in previous research. Li and Ng (2013) examined the causal relationship between economic growth and foreign direct investment (FDI) in South Africa. The annual time series data of real GDP and FDI stocks for South Africa, from 1980 to 2009, were included in the study. First, similar to Li and Hung (2013), an Augmented Dickey-Fuller Test was applied to examine if both variables are stationary. The results showed that they were not stationary. After that, a Johansen Cointegration Test was applied to test if there were cointegrating relationships between these two variables. The results showed that they were not cointegrated, which implied that there was no long-term relationship between real GDP and FDI stocks in South Africa. Despite that fact that these two variables were not cointegrated and non-stationary, the differences were stationary, which could be interpreted as a growth of real GDP and FDI stocks. A VAR model for the growth of FDI stocks and the growth of real GDP was then established. After that, a Granger Causality Test was used to study the causal relationship. The result showed that direction of causality ran from the FDI stocks growth to real GDP growth. Hence, in the short run, the FDI stocks growth could lead to real GDP growth in South Africa. Nevertheless, there was no long run relationship between these factors, as they were not co-integrated.

9.6.1 The unit root test for stationarity

Non-stationary data is often a problem in time series analysis. If the variables are in I(1), it is common to launch a VAR model to examine the relationship and then the Granger causality test. However, VAR only indicates the short-term response of the time series but neglects the long term response. Engle and Granger (1987) suggested that the integrated and non-stationary time series can be cointegrated which implies the time series co-moves in the long run. Thus, an error correction model has been developed to dominate the long-term response among variables. The major advantage of the VECM rather than the VAR model is that VECM better interprets both long- and short-term equations.

This study used Augmented Dickey-Fuller (ADF) test for examining unit root through estimating:

$$\Delta Y_t = \beta_0 + \beta_1 t + \delta Y_{t-1} + \sum_{i=1}^{m} \alpha_i \Delta Y_{t-i} + \varepsilon_t, i = 1, 2, \ldots, m$$

where ε_t denotes white noise error term and $\Delta Y_{t-i} = Y_{t-i} - Y_{t-i-1}$. The null hypothesis is $\delta = 0$ and the alternative hypothesis is $\delta < 0$. Rejecting the null hypothesis indicates that there is no unit root.

We then use the Johansen test to verify the cointegrating relationships among variables. Johansen test concerns m variables can consist $m - 1$ cointegrating relationships. Thus, Johansen (1988a) suggested to modify a generalized ADF to examine the cointegrating relationships among m variables,

$$Y_t = A_t Y_{t-1} + \varepsilon_t$$

where Y_t denotes $m \times 1$ vectors. Thus,

$$\Delta Y_t = (A_t - I) Y_{t-1} + \varepsilon_t$$
$$= \pi Y_{t-1} + \varepsilon_t$$

where Y_t and ε_t are $m \times 1$ vectors while A_t is $m \times m$ matrix of parameters. Generally, if there are r cointegrating relationships within m variables, through checking the significance of the characteristic roots of π, the number of distinct cointegrating vectors can be examined. Since the rank equals to the number of non-zero characteristic roots. For matrix, π, with m eigenvalues, λ_i, in descending order of size,

$$\lambda_1 > \ldots > \lambda_m$$

If the rank is unity, $0 < \lambda_1 < 1$, $\ln(1-\lambda_1) < 0$ and $\ln(1-\lambda_j) = 0, \forall j > 1$. It should be noted that $\lambda_i = 0, \forall i$ for no cointegrating relationship exists.

In Johansen test, there are two test statistics to obtain the number of characteristic roots which are trace test, $\lambda_{trace}(r)$, and lambda-max test, $\lambda_{max}(r, r+1)$,

$$\lambda_{trace}(r) = -T \sum_{i=r+1}^{m} \ln(1-\hat{\lambda}_i)$$

$$\lambda_{max}(r, r+1) = -T \ln(1-\hat{\lambda}_{r+1})$$

where $\hat{\lambda}_i$ denotes the estimated number of characteristic roots obtained from estimated π, and T is the number of observation. The critical values table of both tests are calculated by Johansen and Juselius (1990). If the test statistic is greater than the critical value, it can reject the null hypothesis that there are r cointegrating vectors in favor of the alternative which is at least r + 1 cointegrating relationships.

Finally, assume there are two variables X and Y only, such that:

$$\Delta Y_t = \varphi + \lambda e_{t-1} + \gamma_1 \Delta Y_{t-1} + \ldots + \gamma_i \Delta Y_{t-i} + \omega_1 \Delta X_{t-1} + \ldots + \omega_i \Delta X_{t-i} + \varepsilon_t$$

where ε_t is white noise and $e_{t-1} = Y_{t-1} - \beta X_{t-1}$ is the error term obtained from the cointegration equation. The Granger Causality test infers that X Granger causes Y if the previous periods of variable X has the explanatory power of current Y. We examine the null of $\omega_1 = \ldots = \omega_m = 0$ and $\lambda = 0$ for short run and long run causality respectively. Reject the null means X Granger causes Y.

9.6.2 Research results

The following tables present the results of the unit root test of all the variables in both countries. Assumptions of intercept and trend are included in the test in order to avoid the misleading results of trend stationary (Brook 2008). The lag lengths are selected by using SIC. The results uniformly indicate that all the variables contain unit roots at their level, while they are stationary in their first difference forms.

After confirming all the variables in each country contains unit root at their level and are I(1), the next step is to test whether there is a cointegrating relationship among them. If the answer is positive, VECM is applicable. Johansen cointegration test will be used in this procedure. This test allows for more than one cointegration vectors to be existed. In the case of the Czech Republic, both Trace and Max-Eigen statistic indicate that there is only one cointegrating equation exist among the variables. In the case of South Africa, though the results of trace test and Max-Eigen test are not, one should follow the Max-Eigen test as the final result (Johansen 1988b). Thus, there is only one cointegrating equation in the South Africa case.

The following tables show the results of the Granger causality test in each country. Since this study primarily aims to find out whether housing price plays an important role in monetary transmission mechanism, we are only interested in whether housing prices Granger cause consumption (wealth effect) and investment

Table 9.2 Czech Republic's unit root test results

Variables	Augmented Dickey-Fuller test statistic	
	Level	*First difference*
hpi_c	−0.68	−5.98*
int_c	−3.10	−9.12*
sp_c	−1.45	−4.98*
inv_c	−1.56	−4.57*
cs_c	−0.96	−4.23*

Table 9.3 South Africa's unit root test results

Variables	Augmented Dickey-Fuller test statistic	
	Level	*First difference*
hpi_sa	−1.95	−4.63*
int_sa	−2.10	−4.85*
sp_sa	−1.11	−3.45***
inv_sa	0.15	−3.76**
cs_sa	−2.80	−6.56*

Note: *, ** and *** indicate the rejection of the null hypothesis (contain a unit root) at 1%, 5% and 10% level of significance, respectively. The lag lengths are selected by using SIC.

Table 9.4 Results of Czech Republic's Johansen cointegration test

Null hypothesis: number of cointegrating equations	Trace statistic	Max-Eigen statistic
None	79.94**	35.04**
At most 1	44.90	17.13
At most 2	27.77	14.24
At most 3	13.52	9.30
At most 4	4.22	4.22

Note: ** indicates the rejection of the null hypothesis at 5%.

Table 9.5 Results of South Africa's Johansen cointegration test

Null hypothesis: number of cointegrating equations	Trace statistic	Max-Eigen statistic
None	106.74**	43.96**
At most 1	62.78**	26.38
At most 2	36.40**	23.00
At most 3	13.40	8.56
At most 4	4.84	4.84

Note: ** indicates the rejection of the null hypothesis at 5%.

Table 9.6 Granger causality test results of the Czech Republic

	Short run Granger causality					Long run Granger causality ECT_sa_{it-1}
	Lagged Δhpi_sa	Lagged Δcs_sa	Lagged Δinv_sa	Lagged Δsp_sa	Lagged Δint_sa	
	Chi-sq statistics					t-statistics
Dependent variable						
hpi_c	–	8.36***	1.77	5.17	6.19	−2.26**
cs_c	16.99*	–	10.14**	12.56**	5.86	−3.73*
inv_c	5.68	9.67**	–	11.59**	3.88	−3.09*
sp_c	13.31*	6.36	7.88***	–	2.37	−1.21
int_c	7.88***	10.11**	6.55	8.86	–	2.25**

Note: *, ** and ***indicates the rejection of the null hypothesis at 1%, 5% and 10% level of significance.

(balance sheet effect). In both countries, it is found that housing prices Granger cause consumption in the short run. Further, housing price Granger causes both consumption and investment in the long run. After all, we can discover that there is strong statistical evident that wealth effect and balance sheet effect of housing prices exist in the Czech Republic and South Africa. Thus, housing price plays an important role in the monetary transmission mechanism in both countries.

Table 9. 7 Granger causality test results of South Africa

	Short run Granger causality					Long run Granger causality ECT_c_{it-1}
	Lagged Δhpi_sa	Lagged Δcs_sa	Lagged Δinv_sa	Lagged Δsp_sa	Lagged Δint_c	
	Chi-sq statistics					t-statistics
Dependent variable						
hpi_sa	–	6.10	11.09**	5.40	4.51	–3.05*
cs_sa	14.30*	–	20.84*	10.64**	1.69	–3.60*
inv_sa	1.10	3.40	–	5.02	4.63	–1.88***
sp_sa	5.09	5.10	5.48	–	1.86	–2.62**
int_sa	6.22	26.81*	15.59*	10.17**	–	–1.80***

Note: *, ** and ***indicates the rejection of the null hypothesis at 1%, 5% and 10% level of significance.

9.6.3 Impulse response functions

To figure out the extent of the wealth effect, balance sheet effect and the interest rate exposure of the housing markets in both countries, the impulse response function is chosen as a tool for our analysis. To avoid the causal ordering problem, the generalized impulse response function developed by Pesaran and Shin (1998) was applied.

9.6.3.1 The extent of the wealth effect

Figure 9.2 shows the response of logged real consumption to logged real housing price indices in the Czech Republic and South Africa. The results show that when there is an upward shock of logged real housing prices, the logged real consumption in both countries will increase. This further confirms the existence of the wealth effect created by housing prices fluctuation in both countries. In South Africa, the positive wealth effect shows an inverted U-shaped curve. In the first quarter after the shock, the logged real consumption increases by 0.006531. Then, it starts to increase at an increasing rate until reaching to the peak of 0.0376 in the ninth quarter. After the ninth quarter, the wealth effect starts to wear off. On the contrary, in the Czech Republic, the logged real consumption does not response in the first quarter after the upward shock of logged real housing price. After that, it shows a gentler increase than the case of South Africa. After reaching a peak of 0.27793 in the thirteenth quarter, the positive wealth effect in the Czech Republic started to decline. It can be concluded that the extent of wealth effect created by housing prices fluctuation is greater in South Africa.

9.6.3.2 Extent of the balance sheet effect

Figure 9.3 shows the response of logged real investment to an upward stock of logged real housing price index in both countries. Again, the impulse response

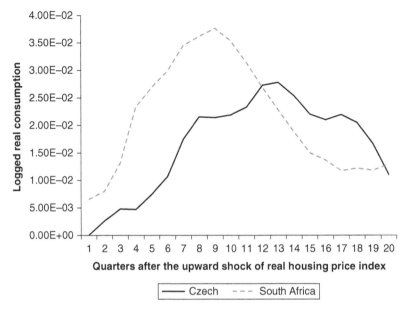

Figure 9.2 Response of logged real consumption to logged real housing price index in Cezch Republic and South Africa

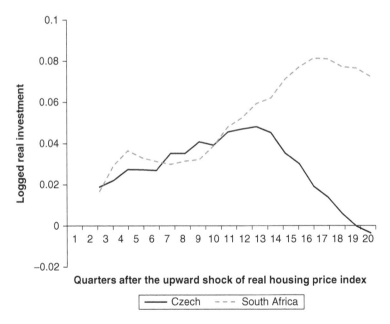

Figure 9.3 Response of logged real investment to logged real housing price index in Cezch Republic and South Africa

function further confirms the existence of the balance sheet effect in both countries. Between the first and the tenth quarter after the upward stock of logged real housing price index, the extent of the positive balance sheet effect in both countries is roughly the same. However, approximately after the tenth quarter, the positive balance sheet effect in South Africa continues to be strengthened while in the Czech Republic, it starts to wear off. Overall, the extent of balance sheet effect, as well as the wealth effect, in South Africa is greater than that in the Czech Republic.

9.6.3.3 Interest rate exposure of the two housing markets

As expected, a rise in interest rate or a tighter monetary policy will lead to a drop in housing price indexes in both countries. However, the extent of a drop in housing price indexes in responding to a rise in interest rate is quite different in the two countries. It is obvious that the housing market in South Africa is more sensitive to upward interest rates than in the Czech Republic. Figure 9.4 presents the impulse response functions base on the VECM. It shows how a shock (an increase) in interest rate, which may be caused by strong upward pressure on global interest rates, affects the housing price index in South Africa and the Czech Republic.

In the Czech Republic, during the first quarter after the upward shock in interest rate, the logged housing price index decreased by 0.006. As time passed, the logged real housing price index started to decrease at an increasing rate between

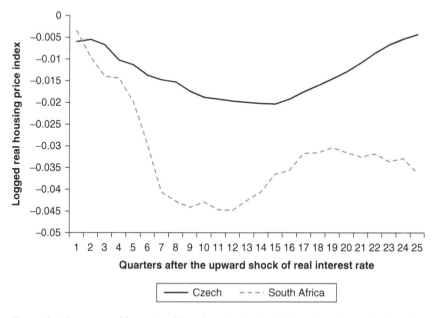

Figure 9.4 Response of logged real housing price index to logged real interest rate in the Czech Republic and South Africa.

the second and the fourteenth quarter after the shock. Nevertheless, starting from the fifteenth quarter, the negative impact on the logged housing price index weakened and tended to disappear. In sharp contrast, although the logged housing price index in South Africa also did not have a vigorous response as it only dropped by 0.0035 during the first quarter, the negative impact of the increased interest rate on the logged real housing price index strengthened rapidly to −0.045 in the twelfth quarter after the shock. Unlike the Czech Republic, the negative impact on housing prices in South Africa did not disappear; although it started to weaken after the twelfth quarter, it continued in the long run.

Based on the abovementioned results of impulse response analysis, we can conclude that when there is an upward shock in interest rate, there will be a significant decrease in housing prices in South Africa. On the other hand, there will be a gentler decrease in housing prices in the Czech Republic. The housing market in South Africa has a greater interest rate exposure.

9.7 Discussion of the results

In summary, the econometric analysis shows that:

1 The extent of both the wealth and balance sheet effects due to housing price fluctuations was greater in South Africa.
2 South Africa faced greater interest rate exposure in the housing market.

These results imply that housing prices in South Africa may drop more than in the Czech Republic when global interest rates rise in the future. Furthermore, as South Africa has a greater wealth and balance sheet effect, a financial crisis may lead to a more serious recession in South Africa. We speculate that the differences in rent deregulation in each country may be the main reason. The value of housing is the sum of discounted future income, i.e. rent. When rental income is regulated by governments, speculation activities in housing markets will also be restricted. It is because the price of a rent regulated house or flat was not determined by the market mechanism given that black market activities are not that serious. In South Africa, the Rental Housing Act [No. 50 of 1999] repealed the rent control (Republic of South Africa Government 1999). Nowadays, housing prices are freely determined and driven by market demand and supply. Thus, foreign capital started to flow into South Africa after rent deregulation and speculation activities increased (Moghadam 2011).

Although rent deregulation was adopted by the Czech Republic in 2006, it was a gentle one and took place from October 2006 to 2012. A housing price bubble is less likely to result in the Czech Republic due to the gentle and late rent deregulation (Zemčík 2008); when rent regulation in South Africa had already been passed, in the Czech Republic it was still a work in progress. About 90% of the population in the Czech Republic still lives in rent regulated housing according to the Global Property Guide (2011). South African has seen more speculation in housing (Moghadam 2011) and, accordingly, a greater interest rate exposure. And both the wealth effect and the balance sheet effect from housing price fluctuations is greater in South Africa.

9.8 Conclusion

With regards to the possible strong upward pressure on global interest rates in the near future, this chapter has studied the interest rate exposure of the housing markets and the role of housing prices in monetary transmission mechanism in the Czech Republic and South Africa. The Granger causality test results indicate that housing price fluctuations will create wealth and balance sheet effect. The results of impulse response functions based on the VECMs show that the extents of both wealth and balance sheet effect are greater in South Africa and South Africa faces a greater interest rate exposure in housing market. The possible reason may be due to less regulation in South African housing, while its lack in the Czech Republic has fueled speculation (Moghadam 2011). Such activities imply a greater wealth and balance sheet effect brought by housing price fluctuations, and greater interest rate exposure. The lesson learnt from this study is that market regulation may help a country to prevent the economy from being hurt seriously by global financial fluctuations. Nevertheless, we do not agree to tighten the rent regulation substantially since too many interruptions from the visible hand may increase the deadweight loss and cause other adverse economic impacts. Nevertheless, some form of regulation may help prevent speculation activities by way of capital gain tax and smoothen the process of public housing supply can be considered.

9.9 References

Becker, C. M. and A. R. Morrison (1999) "Urbanization in Transforming Economies." In *Handbook of Regional and Urban Economics, 1791–1864*.

Brook, C. (2008) *Introductory Econometrics for Finance*. UK: Cambridge University Press.

Chen, M. C. and K. Patel (1998) House Price Dynamics and Granger Causality: An Analysis of Taipei New Dwelling Market. *Journal of the Asian Real Estate Society*, 1, 101–126.

Edelsteina, R. H. and S. K. Lum (2004) House Prices, Wealth Effects, and the Singapore Macroeconomy. *Journal of Housing Economics*, 13, 342–367.

Engle, R. F. and C. W. J. Granger (1987) Co-Integration and Error Correction: Representation, Estimation, and Testing. *Econometrica*, 55, 26.

Fisher, I. (1933) The Debt-Deflation Theory of Great Depressions. *Econometrica*, 1, 337–57.

Gambacorta, L. and P. E. Mistrulli. (2011) *Bank Heterogeneity and Interest Rate Setting: What Lessons Have We Learned Since Lehman Brothers?* Bank for International Settlements.

Global Property Guide (2011) *Czech Republic*.

Hochguerte, S. (2001) The Relation between Financial and Housing Wealth: Evidence from Dutch Households. *Journal of Urban Economics*, 49, 374–403.

Johansen, S. (1988a) Statistical Analysis if Cointegrationg Vectors. *Journal of Economic Dynamics and Control*, 12, 24.

Johansen, S. (1988b) Statistical Analysis of Cointegration Vectors. *Journal of Economic Dynamics and Control,* 12, 231–254.

Johansen, S. and K. Juselius (1990) Maximun Likelihood Estimation and Inference on Cointegration – with Applications to the Demand for Money. *Oxford Bulletin of Economics and Statistics,* 52, 169–210.

Kahn, R. F. (1931) The Relation of Home Investment to Unemployment. *The Economic Journal,* 41, 173–198.

Kiyotaki, N. and J. Moore (1997) Credit Cycles. *Journal of Political Economy*, 105, 211–248.

Leung, C. (2004) Macroeconomics and Housing: A Review of the Literature. *Journal of Housing Economics,* 13, 249–267.

Li, R. Y. M. (2011) "A Statistical Review on the Uneven Impact of Subprime Financial Crisis on Global Property Market." In *Macao International Symposium on Accounting and Finance 2011.* Macau, China.

Li, R. Y. M. (2012) "Chow Test Analysis on Structural Change in New Zealand Housing Price During Global Subprime Financial Crisis." In *18th Annual Pacific Rim Real Estate Society Conference.* Adelaide.

Li, R. Y. M. (2014) *Law, Economics and Finance of the Real Estate Market – A Perspective of Hong Kong and Singapore.* Germany: Springer.

Li, R. Y. M. (2015) Generation X and Y's Demand for Homeownership in Hong Kong. *Pacific Rim Real Estate Journal,* 21, 15–36

Li, R. Y. M. and H. Y. Chan (2013) "The Impact of Eurozone Debt Crisis on China's Property and Land Market." In *European Union Studies Association Asia Pacific Conference.* Macau.

Li, R. Y. M. and K. W. Chau (2011) "The Impact of Institutional Change on Housing Market in China." In *Economy and society in mainland China, Taiwan and Hong Kong: Studies in Entrepreneurship, Finance and Institutions.* Hong Kong: Ovis Press.

Li, R. Y. M. and R. Hung (2013) Rostow's Stages of Growth Model, "Urban Bias" and Sustainable Development in India. *Journal of Contemporary Issues in Business Research,* 2, 170–178.

Li, R. Y. M. and J. Li. (2012) "The Impact of Subprime Financial Crisis on Canada and United States Housing Market and United States Housing Cycle and Economy." In *ICBMG Conference.* Hong Kong.

Li, R. Y. M. and C. Y. Ng (2013) The Chicken-And-Egg Relationship between Foreign Direct Investment Stock and Economic Growth in South Africa. *Journal of Current Issues in Finance, Business and Economics,* 6, 23–38.

Li, R. Y. M. and T. T. Wong (2013) "The Shadow of Eurozone Sovereign Debt Crisis Kept the Eastern Dragon's Property Away from Sunshine? Time Series Analysis on the Negative Externality in China's Property Market by EView 7 Software." In *International Symposium on Computational and Business Intelligence.* New Delhi, India.

Lux, M. (2000) The Housing Policy Changes and Housing Expenditures of Households in the Crech Republic. *Sociological Studies,* 2000, 1–58.

Malpezzi, S. (1999) "Economic Analysis of Housing Markets in Developing and Transition Economies." In *Handbook of Regional and Urban Economics, 1791–1864.*

Mishkin, F. S. (2010) *Economics of Money, Banking and Financial Markets.* Boston: Pearson Addison-Wesley.

Moghadam, R. (2011) Recent Experiences in Managing Capital Inflows – Cross-Cutting Themes and Possible Policy Framework. *IMF Report.*

Movshuk, O. (2010) Balance Sheet Effects on Household Consumption: Evidence from Micro Data. *Hitotsubashi University, the Institute of Economic Research,* 1–38

Nguyen, T. and K. M. Wang (2010) Causality between Housing Returns, Inflation and Economic Growth with Endogenous Breaks. *Journal of the Asian Real Estate Society,* 8, 95–115.

Pesaran, H. H. and Y. Shin (1998) Generalized Impulse Response Analysis in Linear Multivariate Models. *Economics Letters* 58, 17–29.

Pillaya, A. and W. A. Naude (2006) Financing Low-income Housing in South Africa: Borrower Experiences and Perceptions of Banks. *Habitat International* 30, 872–885.

Republic of South Africa Government (1999) Rental Housing Act [No.50 of 1999] Gazette No. 20726.

Royal Oak Interactive (2015) If It Were My Home Compare South Africa to Czech Republic. http://www.ifitweremyhome.com/compare/ZA/CZ

Tobin, J. (1969) A General Equilibrium Approach to Monetary Theory. *Journal of Money, Credit, and Banking*, 1, 15–29.

World Economic Forum (2010) *Global Risk 2010*. Geneva: World Economic forum.

World Economic Forum. (2011) *Global Risk 2011*. Cologny/Geneva.

Zemčík, P. (2008) Is There a Real Estate Bubble in the Czech Republic? *Finance a úvěr-Czech Journal of Economics and Finance*, 61, 49–66.

10 Factors which drive the ups and downs of housing prices in Canada

A Cobb–Douglas approach

*Rita Yi Man Li, Kwong Wing Chau
and Ling Sanchi Li*

10.1 Introduction

In many countries, there is strong enthusiasm for home ownership. In the US, for example, home purchase is considered the American Dream. Home ownership is often identified as an economically efficient action as homeowners have a financial interest in their properties. They are more eager to maintain or enhance the housing value so as to increase their physical and emotional wellbeing as well as their level of satisfaction. Not in the same vein, however, home renters have short-sighted perceptions on local affairs. Many of them fall short on motivation in civic participation and have low social utility. On top of that, home ownership does not only provide a secure place to live, it is also perceived as a means to develop personal wealth which provides financial and economic security (Li 2015).

In Canada, home ownership is important. The rise in average housing prices and a prolonged education deters individuals from home purchase in spite of the low mortgage rate, high employment rate and the government's initiation of a first home purchase scheme. The General Social Survey (GSS) found that most home-buyers are: 1) young adults (aged 25 to 39 years old) living in a rural environment with a permanent job over the past 12 months who did not live with their parents, 2) married couples with children (Turcotte 2008). The Canadian housing market expanded partly due to longer lifespan and rural/urban migration. The influx of immigrants to Canada was another driving force in home purchase (Thomas 2005). New immigrants, however, were less likely to become homeowners due to their temporary jobs (Turcotte 2008).

10.2 Literature review

10.2.1 Inflation and housing prices

Some of the previous research studies conceded that there was a positive correlation between demand shocks and housing prices. They amplified and spread over time. The increase in asset prices raised the debtors' borrowing capacity, allowing them to invest and spend more. The rise in consumer prices reduced the outstanding debt obligations which in turn increased their net worth (Lacoviello 2005). Others found that housing prices caused inflation (Li and Chiang 2012) and housing prices rose even faster than inflation in the US's major metropolitan areas (Case *et al.* 2000). Real estate is also considered as a good financial tool to hedge against inflation (Obereiner and Kurzrock 2012, Li and Ge 2008). Yet, the inflation-hedging ability of housing varies among different countries. It is a good tool in the US in both expected and unexpected inflation. In Australia, most properties hedge against actual and expected inflation but are less effective when there is unexpected inflation. Stevenson and Murray suggested that property could not hedge against expected and unexpected inflation (Li and Ge 2008).

10.2.2 Mortgages, interest rates, money supply and housing prices

Residential mortgages are often considered as the first-order importance for households, financial institutions and macroeconomic stability. The typical household in a developed economy has a dominant liability – a mortgage – as well as one dominant asset – a house. In the modern banking and finance system, financial innovations allow banks to securitize mortgage pools in a relative easy way. As a matter of fact, mortgages are a major fraction of bank assets in modern society (Li 2014b).

As housing prices are quite high in many places around the globe, it is natural that some of us cannot buy a home solely because of lack of wealth. In the light of this, the mortgage market provides a source of funds to homebuyers (Li 2014b) and hence plays an important role in the ups and downs of housing prices.

The relationship between monetary variables and housing price was multi-faceted. The traditional transmission process suggested that there were two way causality between money and housing prices (Goodhart and Hofmann 2008). What, then, affects the money supply in the market? The rises and falls of money supply due to changes in interest rate affected housing prices. In Hong Kong, there was strong correlation between housing prices and interest rates from 1998 to 2001. Reduction in interest rates was strongly correlated to the increase in housing prices until 1997. Nevertheless, the relationship appeared to be non-existent thereafter. While the impact of interest rates tends to be significantly positive in the inflationary pre-1997 period, most of the drop in housing prices since 1998 can be attributed to housing price expectations (Wong *et al.* 2003).

10.2.3 Economics and housing

The macro economy affects housing markets in a number of different ways. The major drivers of housing price includes income (Muellbauer and Murphy 2008) and business cycles (Li 2014a). In Hong Kong, the housing market boomed when the economy was in an expansionary phase, from 1978 to 1981 (Chen *et al.* 2004). In Singapore, housing prices rose substantially between 1980 and 1984 in times of fast economic growth and decreased with the first general recession after independence (Chen *et al.* 2004). In the US, many agree that the housing market downturn caused the economy to stumble (Igan *et al.* 2011, Aspachs-Bracons and Rabanal 2010).

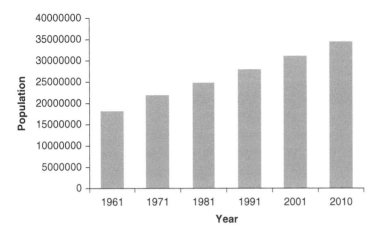

Figure 10.1 Population of Canada from 1961 to 2010 (Statistics Canada 2012)

10.3 Background information about Canada

There are five regions in Canada: Central Canada, the Prairies, the Atlantic region, the West Coast and the North. The population of Canada rose from 18,238,000 in 1961 to 34,378,000 in 2010 but the rate of increase dropped from 30.20% in 1961–1971 to 10.82% in 1991–2010. The growth rate dropped from 1961 to 1991 and became stable in 1991–2001 and in 2001–2010 at 10.67% and 10.82% respectively. This is deal to the drop in population growth. As the birthrate drops and the death rate remains constant, there is a growing ageing problem (Statistics Canada 2012).

10.3.1 Gross Domestic Product (GDP) per capita of Canada

The Gross Domestic Product (GDP) per capita in Canada kept increasing. It rose from US$11,915.81 in 1981 to US$45,888 in 2010 (Nation Master 2012). There was a significant drop in the GDP per capita of Canada in 1998 and 2009 (most of the countries in the world were suffering from the financial crisis during 1997 and 2008) (Conference Board of Canada 2012, Nation Master 2012, CIA World Factbook 2011). From Figure 9.4, Canada's GDP per capita was US$45,888, ranked 11 right after the well-developed countries such as United State and Netherlands.

10.3.2 Inflation rate in Canada

Inflation refers to rises in the general overall price level. As real estate property is one of the many commodities in the market, it is natural to expect that when the prices of everything rise, housing prices should also be on an upswing. Figure 10.3 shows that inflation in Canada was around 2–2.5% every year from 2001 to 2010. In Canada,

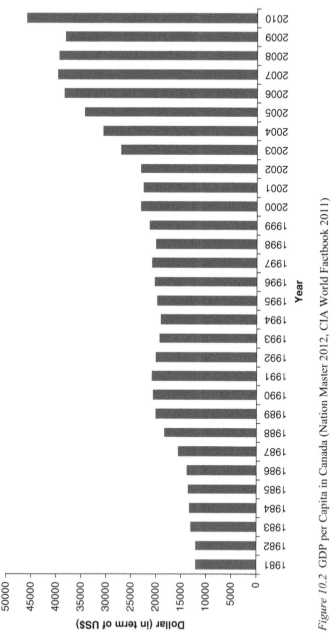

Figure 10.2 GDP per Capita in Canada (Nation Master 2012, CIA World Factbook 2011)

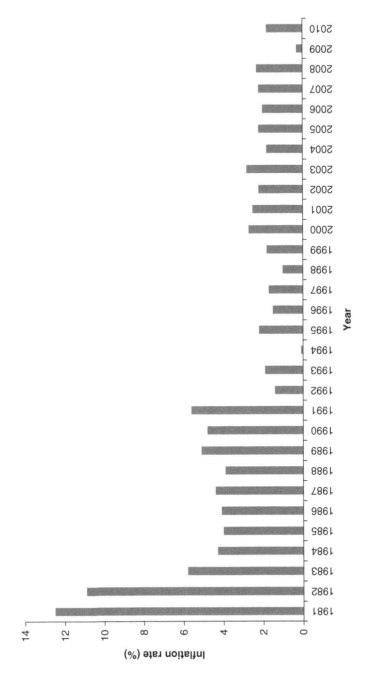

Figure 10.3 Inflation rate (%) in Canada from 1981 to 2010 (Statistics Canada 2012, Bank of Canada 2012a)

inflation was highest in the early 1980s then decreased from 12.5% in 1981 to 1.8% in 2010. There was a significant drop in the inflation rate in 2009 because of the economic downturn during 2008. A significant increase in the inflation rate in 2010 demonstrates that the recovery in the economy of Canada (Bank of Canada 2012b, Statistics Canada 2012, Conference Board of Canada 2012).

10.3.3 Average housing price in Canada

The Toronto Real Estate Board shows the average selling price of single family homes from 1987 to 2010. It is observed that the average housing price in Canada increased steadily, from around $190,000 in 1987 to c.$275,000 in 1989. Then, it decreased to around $205,000 in 1993 due to recession. The average housing price kept increasing afterwards following the recovery in economy. In 2010, the average housing price in Canada for a single family home was $431,463 in Canadian dollars. The average housing price of single family home increased sharply from $198,150 in 1996 to $431,463 in 2010 which was almost 2.2 times as compared to the average housing price in 1996. We predict that the average housing price will keep increasing because of the following reasons (Toronto Real Estate Board 2012):

1   The increase in the demand of housing due to the population growth pushes up the price when the supply of housing remains fairly constant. As a result, the average housing price of a single family house increases (Statistics Canada 2012, NRI Online 2012, Canada Eh 2012).
2   Inflation leads to an increase in the average housing price of a single family home. As the materials used in building a house increase, the average housing price therefore increase (Toronto Real Estate Board 2012).

10.4 Residential mortgage market in Canada

In Canada, people who are over 19 years old can get a mortgage loan. Similar to other countries all over the world, wage affects how much loan can receive (TD Bank Group 2012, Bank of Canada 2012b). Furthermore, the applicants cannot have debt payments exceed 32% of the gross (before tax) income (TD Bank Group 2012). The applicants cannot have been bankrupt in the previous ten years, the previous loan record at the credit bureau affects applicants' opportunities in getting a mortgage loan. After the financial crisis, the banks in Canada reduced the maximum amortization from 35 years to 30 years for mortgage policy, and increased the minimum down payment from 5% of the property value to 10%. These two changes have tightened mortgage policy and protected banks with more securities (TD Bank Group 2012, Bank of Canada 2012b). These six major banks in Canada – Bank of Montreal, Bank of Nova Scotia, Banque Nationale, Canadian Imperial Bank of Commerce, Royal Bank Financial Group, and TD Bank Financial Group – are collectively known as "The Big Six." They all offered both fixed and variable rate mortgages.

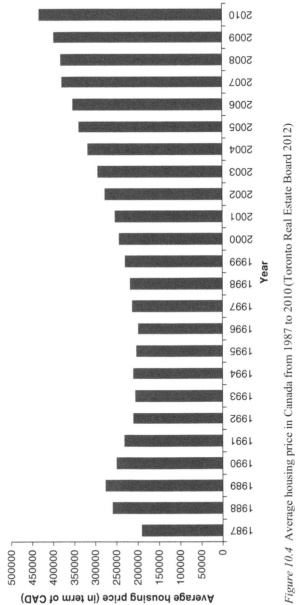

Figure 10.4 Average housing price in Canada from 1987 to 2010 (Toronto Real Estate Board 2012)

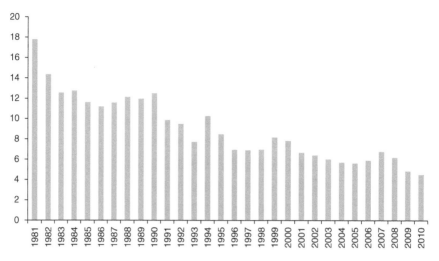

Figure 10.5 Average residential mortgage lending rate (5 years) in Canada from 1981 to 2010 (Bank of Canada 2012b)

The average mortgage rate in Canada dropped from 17.79% in 1981 to 4.5% in 2010. Although it increased from 5.6% in 2005 to 6.75% in 2007, it dropped again afterwards (Bank of Canada 2012b). Furthermore, the busiest months for mortgage applications were the first quarter with 30.5% and the fourth quarter marked the lowest portion of mortgage applications (Canequity 2012).

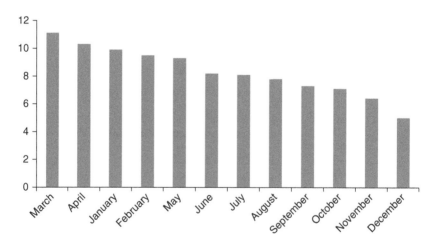

Figure 10.6 Busy months for Canadian mortgage applications (Canequity 2012)

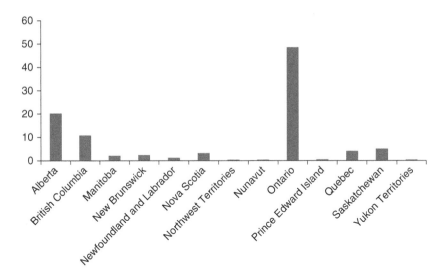

Figure 10.7 Mortgage applications compared to all other provinces (Canequity 2012)

Among all the provinces in Canada, Ontario had received the largest percentage of applications at 48.43%, followed by Alberta (20.09%) and British Columbia (10.07%). The Yukon Territories, Northwest Territories and Nunavut recorded the lowest rate of 0.34%, 0.33% and 0.09% respectively (Canequity 2012). The statistical data also displayed an obvious positive correlation between population density and the number of mortgage applications. For example, Toronto in Ontario and Calgary in Alberta are the cities with high population density. These two places also recorded the highest percentage of mortgage applications. In Calgary, the percentage of mortgage application was 7.002% Applicants on average aged 37 received $66,455.83 mortgage loan. In Toronto, the percentage of mortgage applications was 4.796% and the applicants for mortgage were at the average age of 37, received an average income of $72,660.43 and the average loan of $260,402.74 (Canequity 2012). Yukon Territories, Northwest Territories and Nunavut are provinces with low density of population recorded the lowest percentage of mortgage applications so as the home purchase (Canequity 2012, Statistics Canada 2012). Furthermore, Canadians are moving from rural homes to more urban areas; rural urban migration creates a huge demand on urban areas' housing markets and on mortgages. That offers a vivid explanation of why large cities like Calgary and Toronto have more mortgage applications than in in rural areas, such as the cities and towns in the province of Nunavut (Thomas 2005).

The most popular mortgage application type was Qualification, accounting for 24% of mortgage applications, followed by first-time Buyer and Refinance

Table 10.1 Top producing cities/towns for mortgages within Canada (Canequity 2012)

	City or town within Canada	Percentage of mortgage applications	Average age	Average income	Average co-income	Average amount of loan
001:	Calgary, AB	7.002%	37	$66,455.83	$45,567.96	$225,596.80
002:	Toronto, ON	4.796%	37	$72,660.43	$52,558.79	$260,402.74
003:	Edmonton, AB	4.534%	36	$58,646.92	$42,490.21	$206,098.93
004:	Mississauga, ON	2.548%	38	$65,538.97	$43,755.27	$227,840.17
005:	Ottawa, ON	2.375%	37	$64,955.51	$48,624.08	$203,680.97
006:	Regina, SK	1.875%	33	$47,368.38	$36,425.90	$130,134.02
007:	London, ON	1.566%	36	$51,968.40	$39,991.30	$154,783.79
008:	Brampton, ON	1.523%	38	$56,138.25	$40,525.81	$193,597.23
009:	Hamilton, ON	1.486%	37	$51,602.98	$36,776.69	$147,218.17
010:	Winnipeg, MB	1.159%	37	$50,243.66	$35,877.45	$136,789.05
011:	Kitchener, ON	1.072%	37	$56,416.53	$45,345.35	$176,087.75
012:	Vancouver, BC	1.013%	38	$67,036.25	$43,487.19	$274,374.22
013:	Scarborough, ON	0.883%	37	$55,186.88	$39,536.81	$219,214.66
014:	Montreal, QC	0.873%	38	$65,333.75	$44,990.90	$221,483.61
015:	Barrie, ON	0.844%	36	$55,422.09	$36,430.93	$168,816.30
016:	Windsor, ON	0.813%	36	$56,059.76	$40,722.34	$127,295.46
017:	Burlington, ON	0.807%	40	$70,135.40	$44,784.88	$199,047.96
018:	Saskatoon, SK	0.753%	36	$52,542.44	$40,860.57	$165,225.86
019:	Surrey, BC	0.733%	39	$62,578.10	$44,998.96	$218,982.55
020:	Victoria, BC	0.694%	40	$59,686.09	$41,621.49	$229,634.63
021:	Oakville, ON	0.692%	40	$76,973.90	$47,771.29	$222,197.84
022:	Guelph, ON	0.642%	36	$57,785.87	$41,063.20	$195,220.76
023:	Cambridge, ON	0.634%	37	$56,784.73	$39,524.82	$175,881.20
024:	Oshawa, ON	0.630%	38	$54,647.29	$41,268.40	$149,341.64
025:	Red Deer, AB	0.622%	34	$58,444.40	$39,168.80	$203,669.32
026:	Fort Mcmurray, AB	0.585%	35	$94,521.82	$58,277.04	$322,597.96
027:	St Catharines, ON	0.541%	36	$49,567.00	$36,294.63	$150,733.57
028:	Richmond Hill, ON	0.525%	40	$70,397.67	$55,758.91	$215,600.22
029:	Etobicoke, ON	0.525%	37	$55,773.80	$42,106.29	$221,997.99
030:	Brantford, ON	0.523%	36	$49,314.27	$33,804.65	$147,762.80

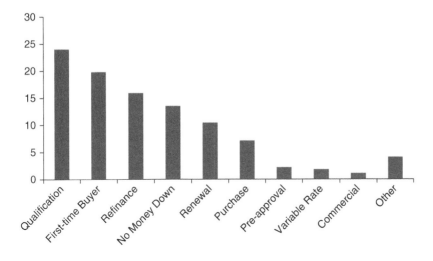

Figure 10.8 Product popularity breakdown for all of Canada (Canequity 2012)

at 19.8% and 15.9% respectively. Furthermore, there were other mortgage applications types, including No Money Down, Renewal, Purchase, Pre-approval, Variable Rate, Commercial and Other. As marriage is one of the important reasons for home purchase among the younger generation (Li 2015), almost half of the applicants are married (49%), followed by singles (28.3%) (Canequity 2012).

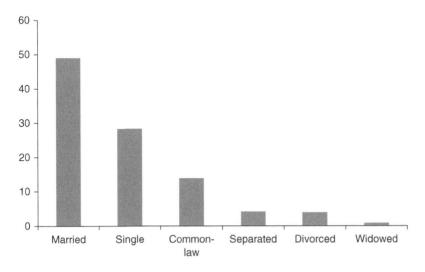

Figure 10.9 Marital status of mortgage applications (Canequity 2012)

10.5 Research method and results

The Cobb–Douglas specification is a multiplicative form. Since it is linear in logs, we can use normal linear approximation techniques (Brierly and Feiock 1993). The major advantage is easy to analyze, and it appears to be a good approximation to actual productions (Yuan *et al.* 2009). It has been widely adopted in energy market analysis (Wei 2007) and information technology (Pendharkar *et al.* 2008). Dillon (2015) adopted the Cobb–Douglas utility function to analyze the impact of the congestion and cordon pricing on rents within the city. Houshyar *et al.* (2015) adopted the Cobb–Douglas model to study the inputs and output relationship between the energy consumption patterns of tomato production and the corresponding GHG emissions in the central Fars province, the main tomato producer region in Southwest Iran.

In this chapter, we considered the housing prices to be the output which was affected by the inputs of inflation rate, GDP and mortgage. And we adopted the Cobb–Douglas production function as follows:

$$Y(HOUS) = A \, (INFLA)^\alpha \, B(GDP)^\beta \, C(MOR)^\gamma$$

This equation equation suggests that inflation, GDP and mortgage rate are the important input factors which will eventually affect the output, i.e. housing price. Furthermore, if

$$\alpha + \beta + \gamma = 1$$

Returns to scale[1] are constant, then:

$$\alpha + \beta + \gamma < 1$$

Returns to scale are decreasing, and if:

$$\alpha + \beta + \gamma > 1$$

Then, returns to scale are increasing (Tan 2008). Generally speaking, the distributions of the data we use are known only asymptotically, i.e. the data does not have the correct size, and inferential comparisons and judgments based on them could be misleading. Data as such from a direct residual resampling gives:

$$Y^* = A\alpha^* + \beta^*$$

where β^* are i.i.d observations $\beta 1^*; \ldots; \beta t^*$, drawn from the empirical distributions $(\hat{F}_\beta)$, putting mass $1/T$ to the adjusted OLS residuals:

$$(\hat{\beta} + \beta^*), i = 1, \ldots, T$$

---

1 Returns to scale refers to a technical property of production which examines the changes in output subsequent to a proportional change in all inputs (Tan 2008).

Table 10.2 Summary statistics of bootstrap data using the observations 1–999 for the variables (999 valid observations)

	LOGHOUS	*LOGGDP*	*Log IF*	*LOGMOR*
Mean	−0.99360	0.94439	3.6010	5.8901
Median	−0.99382	0.94547	3.7013	5.9675
Minimum	−1.0980	0.84401	−0.46545	3.0916
Maximum	−0.90057	1.0416	4.1629	6.4188
Standard deviation	0.032929	0.033259	0.41012	0.35743
C.V.	0.033141	0.035217	0.11389	0.060683
Skewness	−0.14587	−0.095310	−2.7830	−2.3598
Ex. kurtosis	−0.13424	−0.22675	15.529	10.414

The basic principle of the Bootstrap is to draw a number of Bootstrap samples from the model under the null hypothesis, calculate the Bootstrap test statistic. The Bootstrap test statistic (Ts*) can then be calculated by repeating this step N times. After that, we take the ath quintile of the bootstrap distribution of T*s and obtain the a-level "bootstrap critical values" b*. We then calculate the test statistic (Ts) which is the estimated test statistic using the actual data set (Salman and Shukur 2004). Efron (1979)'s paper concedes that one of the possible application of Bootstrap is data expansion. It has been used widely even though there are already many other resampling techniques for dependent data (Kreiss and Paparoditis 2011). Bootstrapping can also be used to estimate the sampling distribution of a statistic by doing repeated sampling (i.e. resampling) with replacement from an original sample (Novoa and Mendez 2009).

We used the yearly inflation rate, GDP and mortgage rate data from 1987 to 2011 in Canada as input variables to test their impact on output value, i.e. housing price (the reason for choosing them is based on the data limitation problem, we cannot find other relevant variable which is longer than 1978 through 2011). As a small number of data points violated the error distribution requirement, we first enlarged the number of data points by the bootstrap procedure with 1,000 replications and then took the log and put the data to the linear Cobb–Douglas model. The results showed that there was negative significant relationship between change in housing price (LOGHOUS) and inflation (LOGIF), which refuted the previous research articles' findings. We speculated that inflation reduced home-buyers' purchasing ability: Canadians waited for the fall of housing prices to buy new ones. Furthermore, there was no significant relationship between housing price and GDP (LOGGDP) and mortgage (LOGMOR). The sum of coefficient of indicated returns to scale were decreasing if:

LOG GDP + LOGIF+ LOGMOR < 1

10.6 Conclusion

Canada is a country with a very strong economy and housing markets. This chapter has studied the impact of various macroeconomic factors on the housing prices in

Table 10.3 Linear Cobb–Douglas result

Model 1: OLS, using observations 1–999 (n = 826)
Missing or incomplete observations dropped: 173
Dependent variable: LOGHOUS

	Coefficient	Std. error	t-Ratio	p-Value
const	−0.961588	0.0379064	−25.37	2.59e-105 ***
LOGINF	−0.00563155	0.00275204	−2.046	0.0410 **
LOGGDP	−0.00319018	0.0342076	−0.09326	0.9257
LOGMOR	−0.00156273	0.00316463	−0.4938	0.6216
	Mean dependent var	−0.994077	S.D. dependent var	0.032615
	Sum squared resid	0.872828	S.E. of regression	0.032586
	R-squared	0.005431	Adjusted R-squared	0.001801
	F(3, 822)	1.496205	P-value(F)	0.214160
	Log-likelihood	1658.085	Akaike criterion	−3308.171
	Schwarz criterion	−3289.304	Hannan-Quinn	−3300.934

Excluding the constant p-value was highest for variable 8 (LOGGDP)

** significant at 95% level
*** significant at 99% level

Canada. It proposed the use of the Cobb–Douglas Model to estimate input factors, including annual GDP, inflation rate and mortgage rate on output housing price from 1978 to 2011 in Canada. As there was difficulty in searching for long historical data in some cases for the model, a bootstrap technique was used to expand the data. It was found that there was an insignificant relationship between housing price, GDP and mortgage rate. Nevertheless, a negative relationship between inflation and housing price was shown in the research results.

Acknowledgement

An earlier version of this chapter has been presented in Global Chinese Real Estate Congress 2012 (GCREC). The author would like to thank the conference organizer for granting the right to republish with revision and extension.

10.7 References

Aspachs-Bracons, O. and P. Rabanal (2010) The Drivers of Housing Cycles in Spain. SERIEs: *Journal of the Spanish Economic Association*, 1, 101–130.

Bank of Canada (2012a) Inflation Calculator. http://www.bankofcanada.ca/rates/related/inflation-calculator/

Bank of Canada (2012b) Rates and Statistics. http://www.bankofcanada.ca/rates/

Brierly, A. B. and R. C. Feiock (1993) Accounting for State Economic Growth: A Production Function Approach. *Political Research Quarterly*, 46, 657–669.

Canada Eh (2012) *Canada Eh?*

Canequity (2012). Mortgage Statistics for Canada. https://www.superbrokers.ca/stats/canadian-mortgages/

Case, K. E., E. L. Glaeser and J. A. Parker (2000) Real Estate and the Macroeconomy. *Brookings Papers on Economic Activity*, 119–162.

Chen, M. C., Y. Kawaguchi and K. Patel (2004) An Analysis of the Trends and Cyclical Behaviours of House Prices in the Asian Markets. *Journal of Property Investment and Finance*, 22, 55–75.

CIA World Factbook (2011). Canada Economy 2011. https://www.cia.gov/library/publications/the-world-factbook/geos/ca.html

Conference Board of Canada (2012). Income per capita (GDP per capita) Country Ranking. http://www.conferenceboard.ca/hcp/details/economy/income-per-capita.aspx

Dillon, H. S. (2015) Social Optimality of Cordon Area Congestion Pricing in a Monocentric City. *Jurnal Perencanaan Wilayah dan Kota* 26, 43–53.

Efron, B. (1979) Bootstrap Methods: Another Look at the Jackknife. *The Annals of Statistics* 7, 1–26.

Goodhart, C. and B. Hofmann (2008) House prices, Money, Credit, and the Macroeconomy. *Oxford Review of Economic Policy*, 24, 180–205.

Houshyar, E., T. Dalgaardb, M. H. Tarazkar and U. Jørgensen (2015) Energy Input for Tomato Production What Economy Says, and What Is Good for the Environment *Journal of Cleaner Production*, 89, 99–109.

Igan, D., A. Kabundi, F. N. De Simone, M. Pinheiro and N. Tamirisa (2011) Housing, Credit, and Real Activity Cycles: Characteristics and Comovement. *Journal of Housing Economics,* 20, 210–231.

Kreiss, J. P. and E. Paparoditis (2011) Bootstrap Methods for Dependent Data: A Review. *Journal of the Korean Statistical Society,* 40, 357–378.

Lacoviello, M. (2005) House Prices, Borrowing Constraints, and Monetary Policy in the Business Cycle. *The American Economic Review,* 95, 739–764.

Li, J. and Y. H. Chiang (2012) What Pushes Up China's Real Estate Price? *International Journal of Housing Markets and Analysis,* 5, 3–22.

Li, L. H. and C. L. Ge (2008) Inflation and Housing Market in Shanghai. *Property Management,* 26, 273–288.

Li, R. Y. M. (2014a). *Law, Economics and Finance of the Real Estate Market – A Perspective of Hong Kong and Singapore*. Germany: Springer.

Li, R. Y. M. (2014b) Mortgage Regulations Changes in the US, China and Australia from 20th Century Through 2011. *Real Estate Finance,* 31, 84–94.

Li, R. Y. M. (2015) Generation X and Y's Demand for Homeownership in Hong Kong. *Pacific Rim Real Estate Journal,* 21(1), 15–36.

Muellbauer, J. and A. Murphy (2008) Housing Markets and the Economy: The Assessment. *Oxford Review of Economic Policy,* 24, 1–33.

Nation Master (2012). Canada Economy. http://www.nationmaster.com/country-info/profiles/Canada/Economy

Novoa, C. M. and F. Mendez (2009) Bootstrap Methods for Analyzing Time Studies and Input Data for Simulations. *International Journal of Productivity and Performance Management*, 58, 460–479.

NRI Online. (2012). Welcome2Canada – Medical Insurance For Canada.

Obereiner, D. and B. M. Kurzrock (2012) Inflation-hedging Properties of Indirect Real Estate Investments in Germany. *Journal of Property Investment and Finance,* 30 218–240.

Pendharkar, P. C., J. A. Rodger and G. H. Subramanian (2008) An Empirical Study of the Cobb–Douglas Production Function Properties of Software Development Effort. *Information and Software Technology,* 50, 1181–1188.

Salman, A. K. and G. Shukur (2004) Testing for Granger Causality Between Industrial Output and CPI in the Presence of Regime Shift: Swedish Data. *Journal of Economic Studies,* 31, 492–499.

Statistics Canada. (2012). Consumer Price Index, Historical Summary http://www.statcan.gc.ca/tables-tableaux/sum-som/l01/cst01/econ46a-eng.htm

Tan, B. H. (2008). Cobb–Douglas Production Function. http://docentes.fe.unl.pt/~jamador/Macro/Cobb–Douglas.pdf

TD Bank Group. (2012). Our Business. http://www.tdbank.com/

The Daily (2007) *2006 Census: Families, Marital Status, Households and Dwelling Characteristics.*

Thomas, D. (2005) Socio-Demographic Factors in the Current Housing Market. *Canadian Economic Observer.*

Toronto Real Estate Board. (2012). TREB (Toronto Real Estate Board) Average Single Family Historical Home Prices and trends for Toronto and Mississauga.

Turcotte, M. (2008). Canada Social Trend: Young People's Access to Home Ownership.

Wei, T. (2007) Impact of Energy Efficiency Gains on Output and Energy Use with Cobb–Douglas Production Function. *Energy Policy,* 35, 2023–2030.

Wong, T. Y. J., C. M. E. Hui and W. Seabrooke (2003) The Impact of Interest Rates Upon Housing Prices: An Empirical Study of Hong Kong's Market. *Property Management,* 21, 153–170.

Yuan, C., S. Liu and J. Wu (2009) Research on Energy-saving Effect of Technological Progress Based on Cobb–Douglas Production Function. *Energy Policy,* 37, 2842–2846.

11 An econometric analysis on REITs cycles in Hong Kong, Japan, the US and the UK

Rita Yi Man Li and Kwong Wing Chau

11.1 Introduction

A lot of investors want to invest in real estate markets, which are traditionally considered as local businesses; for example, houses, industrial buildings and hotels. Nevertheless, expensive housing often bars the entry of many investors. Problems such as lumpiness, obsolescence, liquidity, poor quality due to asymmetric information (Li 2012, Li 2009, Li 2010a, Li 2010b), short-selling constraints, indivisible and high taxation and transaction costs deter many investors from engaging in direct real estate investments (Li and Chow 2015, Cotter and Roll 2015).

Authorized by the Congress in 1960, REITs are securitized, indirect public real estate. They are listed in stock markets that offer small investors golden opportunities to invest in real estate markets which could otherwise be too expensive for them (Wu *et al.* 2011). Previous research reveals that REITs offer the opportunity to observe the possible changes in residential real estate prices (Cotter and Roll 2015).

REITs are prohibited from managing properties actively and receive income from rental and interest payments only (Brady and Conlin 2004). As REITs are traded on the stock market, they provide a foreseeable persistent income stream (all except Turkish REITs provide dividends to investors), efficient investment opportunities, and favorable hedging opportunities to investors (Ambrose *et al.* 2005, Erol and Tirtiroglu 2011), more efficient than the direct real estate markets in general (Li and Li 2011), they are attractive to investors who wish to invest in real estate markets without sacrificing liquidity. Although REITs are technically inefficient (Anderson *et al.* 2002), REITs companies enjoy exemption from corporate income tax and savings in operating expenses by foregoing costly tax-minimizing strategies (Tang and Jang 2008). They developed rapidly by acquiring well-performed income generating properties from partnerships and individuals in the late twentieth century (Ambrose and Linneman 1999, Lee and Pai 2010, Brady and Conlin 2004).

In general, REITs can be classified into index and non-index REITs. Index REITs usually include some domestics REITs, which are not equally-weighted. Some REITs are excluded because of their relatively small size (Ambrose *et al.* 2007).

Alternatively, REITs can be classified into equity, mortgage and hybrid REITs: equity REITs invest and run commercial property, mortgage REITs invest in mortgage backed securities and commercial real estate mortgages, hybrid REITs invest in properties and mortgages. The equity REIT comprises approximately 92% of the total REIT market capitalization (Devaney 2012).

11.2 Factors which drive the REITs prices

Similar to other assets, REITs' prices are estimated according to the present values of a series of expected future cash flows, i.e. series of expected returns are divided by the interest rate. The asset prices reflect time variation in expected returns and/or potential growth in cash flows. As REITs are affected by market fundamentals they can be classified into: 1) rent rate, occupancy rate (Chiang 2015), and operating expenses prior industry level returns (Ling and Naranjo 2006), and 2) expected returns (Chiang 2015), such as long- and short-term interest rate changes (Allen *et al.* 2000), risk tolerance and other discount factors. This grouping is in line with the financial economics literature about the behavior of common stock prices: stock prices are normalized by dividends (Chiang 2015).

Alias and Soi Tho (2011) analyzed the performance of six REITs in the UK and Malaysia. Their findings showed that total revenue was a major factor which affected the performance of the largest M-REITs and UK-REITs. Furthermore, the profit margin of the REITs was raised by approximately 9% for every one billion increase in market capitalization. Furthermore, institutional investors herd in and out of REITs according to the performance of the underlying commercial real estate assets (Das, Freybote and Marcato 2014). Hence, it is often observed that the ups and downs of physical asset prices also affect the REITs markets.

Newell and Peng (2009) investigated the change in risk profile and portfolio diversification benefits of A-REITs over 1996–2008. High gearing was seen as the most critical factor in A-REIT under-performance during the global subprime financial crisis. In other research on Asia REITs markets, it was found that the existence of cross-border diversification opportunities remain even though the markets were cointegrated since the global subprime financial crisis. Short-run causality tests illustrate that the number of causality relationships drop over time. The results prove that domestic REIT investors can diversify the benefits by incorporating some international REITs to domestic portfolio but they need to review their portfolios from time to time since the correlations among markets could change (Loo *et al.* 2015).

Bers and Springer (1997) showed that over-sized REITs might become more efficient by downsizing, spinning off assets or grouping the assets of over-sized REITs into more efficient subgroups. Hayunga and Stephens (2009) found that dividend payments pushed up the prices of REITs. McCue and Kling (1994)'s research suggested that the macroeconomy explained 60% of the changes in Equity REITs prices. Among these macroeconomic variables, nominal interest rates contributed the greatest percentage of the changes.

Ghosh *et al.* (1998) found that a portfolio of 69 REITs reacted negatively to announcements of poor performance real estate portfolios during the real estate crisis of 1989–91 in the US. Ewing and Payne (2005) suggested that shocks due to inflation, economic growth, and monetary policy lead to lower than expected returns in REITs but a shock to the default risk premium was associated with higher future returns. Similar to physical real estate markets, which were affected by the financial crisis, the US residential REITs were very volatile in the global subprime financial crisis from January 2007 to February 2009. Residential REITs fell 53% and recovered 90% from February 2009 to December 2011. Furthermore, industrial REITs decreased by only 40% but recovered 31% of its value from February 2009 to December 2011 only (Sun *et al.* 2015). In Australia, returns from REITs consistently outperformed the general equities. Nevertheless, the sector lost more than 50% of its market value during the Global Financial Crisis. The devaluations were due to the high levels of borrowing during the sector's boom phase 2001 through 2007 (Yong and Singh 2015). Glascock *et al.* (2000) found that REITs were cointegrated with the bond market before 1992 but vanished afterwards.

Adams *et al.* (2015) suggested that risk spillovers from changes in economic factors may come from local economic shocks which propagate to neighboring regions instead of national economic shocks. For example, an adverse shock in a concentrated industry such as the automobile industry in Detroit is relevant for REITs which own properties in that region. When regional economic shock information diffuses to the market, investors will anticipate the future layoffs in service sector and the possible residential properties' rental income losses. Triggered by the spillover effects from retail and industrial REITs, there will then be an increase in selling pressure on residential and office REITs. This is because income generating residential stock constitutes denser multifamily houses that are usually located in the central area of a metropolitan city. The closing of car factories and the subsequent decrease in purchasing power from the layoffs in related supporting industries will have a direct impact on industrial and retail REITs. Some workers will move to other regions for work thereby raising the vacant residential properties supply which lead to spillovers effect from one type of REIT to another.

Furthermore, investors, speculators and researchers have long been interested in understanding the interrelations between stock markets. Theoretical models such as International Capital Asset Pricing Model (ICAPM), generalization of the domestic CAPM and arbitrage pricing theory (ABT) have extended their arms to cover the international financial markets application (Hecq *et al.* 2000). There are many academic articles about various stock cycles in different geographical areas; for example, Kamstra *et al.* (2003) studied the relationship between various cycles in the northern and southern hemispheres.

The above literature suggested that research on REITs cycles is scarce. Hence, we attempt to fill this gap by studying the relationship between various REITs cycles. The chapter is organized as follows. It first provides a brief introduction on REITs and property stocks. After that, it sheds light on the diversification function of the REITs from different angles, including geographical and trades etc. It then analyze previous research on cycles and provides details about the data used. Finally, it throws light on the method of research and the result of cycles studies in four places.

11.3 REITs and property stocks

Many REITs investors are of the view that REITs and property stocks are similar as they are sold and bought on stock markets. Some authors – Chen and Tsang (1988), Allen *et al.* (2000), Jirasakuldech *et al.* (2006) – have even renamed REITs "REITs stocks," despite the obvious differences between the two. Similar to studies of correlation among international housing markets (Li and Li 2012), one of the major strands of the literature has focused on the relationship between REITs and property stocks.

Compared to the listed property stocks, REITs are exempted from corporate income tax; they are required to distribute the majority of their net income to shareholders. Second, REITs are characterized by low level gearing and leverage ratios. Third, REITs are seldom involved in construction and development activities. Their revenue mainly comes from rents or interest payments on mortgage loans, which are relatively stable across business cycles. Some listed property companies, especially in the Hong Kong SAR and Japan, tend to have other lines of business such as development activities, telecommunications and shipping (Zhu 2002).

Oppenheimer and Grissom (1998)'s Spectra results show significant co-movement between US REITs and stock indices. Glascock *et al.* (2000) found that REITs were segmented from the common stock market from 1972 to 1991. However, from 1992 to 1996, we observed cointegration existed between stocks and REITs. According to Tobias *et al.* (2009), REITs were more risky than utility stocks. Others compared the risky nature of REITs and stocks (Pagliari and Webb 1995, Stevenson 2001, Giliberto 1990).

A key feature, which differentiates REITs from property stocks, is the minimum mandatory dividend payment in most jurisdictions. This high yield nature makes REITs more sensitive to interest rates in discounting future expected dividends. Moreover, the coupon-like nature and high dividends of the underlying rental income make REITs share some characteristics with fixed-income securities (Akimov *et al.* 2015).

11.4 Diversification function of REITs

In investment science, it is commonly known that well-diversified portfolios lower unsystematic or idiosyncratic risk (Hui and Chan 2012, Abugri and Dutta 2014). The modern portfolio theory and its extensions, and low international correlation across markets, provide grounds for global portfolio diversification (Schindler 2009). REITs are often regarded as a good risk diversification instrument in portfolio asset allocation (Ahmadi *et al.* 2010) and a bond substitute

(Reynolds 1997). For example, although index and non-index REITs share a lot of similarities, there is no co-movement relationship between them (Ambrose *et al.* 2007). Therefore, a combination of the two provides a golden opportunity for investment diversification. However the diversification benefit of REITs as mixed asset portfolios was weakened during the 2008 global financial crisis (Newell *et al.*, 2010).

Glascock *et al.* (2000) investigated the relationship between equity REITs and mortgage REITs. They found that they diverged after 1992. Previously, there had been a two-way causality relationship between the two types; however, after the sub-prime crisis of 2008, Equity Real Estate Trust funds outperformed the Dow Jones Equity REIT (Ahmadi *et al.* 2010). Newell *et al.* (2010) found that HK-REITs had strong risk-adjusted returns but reduced in diversification benefits from 2005 to 2008. Chen *et al.* (2005) stated that, although REITs did not function well in portfolio diversification before 1985, they improved on the mean-variance frontier between 1986 and 2002.

Li and Yung (2007)'s GARCH (1, 1), EGARCH, and GARCH-M models showed that there were significant volatility transmissions between the Atlantic and Pacific regions. Fei, Ding and Deng (2008) found that the time-varying correlations of REITs were caused by historical stock returns and macroeconomic variables.

Another strand of the literature focused on regional and global contagion in REITs during financial crises. Tsai *et al.* (2012) found a contagion effect among REITs in the US, Canada, Australia and most of the European REITs market. However, Asian REITs (except Singapore) received no contagion from the US REITs market.

Kim *et al.* (2002)'s research showed that the hotel REIT underperformed industrial and office REITs; thus, diversification functions existed. Scott *et al.* (2005)'s research found that REITs primarily dealing with hotel properties had significantly lower returns than self-storage property REITs. Retail was significantly negatively related to Jensen's Alpha. Health care and office were significantly negatively related to Sharpe's measure. With the help of dynamic Copula models, Ibbotson (1990) found that the correlation between REITs and the direct real estate market were weaker than that between REITs and equity markets. Compared to the direct real estate market, REITs outperformed the former as there was no good quality information on price, due to confidentiality and lack of trade in the direct real estate market (Geltner *et al.* 2003).

11.5 Business cycle

Economic change is a law of life. We usually associate economic instability with business recessions and expansions, and we are accustomed to expansion, downturn, contraction and upturn of economic fortunes as the business cycle (Burns 1969).

Previous research showed that economic fluctuations like business cycle often affect welfare, by affecting the consumption growth rate (Barlevy 2004). Kodama

(2013) suggested that the cost of business cycles can be measured by an index representing the welfare loss originated from business cycles.

11.6 Direct real estate cycle analyses

Research on real estate cycles started by Homer Hoyt in 1933 (Mueller 2002). Since then, many results suggested that every city has its own real estate cycle characterized by different lengths and degree of change, affected by demand and supply for local space (Mueller 2002). For example, direct real estate market studies cyclical relationship has been studied extensively (Galiniene *et al.* 2006, Renaud 1997, Gordon *et al.* 1996, Kaiser 1997, Mueller 1999, Mueller 2002, Wernecke *et al.* 2004). In the US, the housing price cycle of the 2000s was characterized by large price increases followed by steep declines and appear to have very different supply elasticities (Davidoff 2013). In Spain, an increase in the proportion of part-time jobs over total employment can exhibit, a deeper business cycle and deeper housing cycle, ceteris paribus. This phenomenon could be more relevant if the part-time jobs are popular among the young adults, who have found this type of jobs as their first working experience (Arestis 2015). Furthermore, previous research also noted that financial integration amplified the housing cycles by strengthening the spillover from residential market to the rest of the economy (Loutskina and Strahan 2015).

Previous research showed that the sales volume could drop by 50% or more from peak to trough in a housing cycle. Even though the most dramatic examples were recorded in local markets, it was evidenced that there were strong positive correlation between trading volumes and aggregate prices at national level in the US, the UK and France. In the boom phase, houses were sold quickly at prices close to or even many times above the home sellers' asking prices. In the bust phase, however, homeowners needed to wait for a long time with asking prices were set well-above the expected selling prices. Hence, many sellers eventually withdrew their properties without selling out (Genesove and Mayer 2001).

There was, however, scarce research about the length and amplitude of cycles in different places (Reed and Wu 2010). Understanding cyclical activity is important in efficient portfolio asset management. When property appraisal does not take cyclical fluctuations into account may produce unrealistic valuation, resulting in poor choice of property assets selection or removal from investment portfolio (Wilson and Okunev 1999).

11.7 REITs cycles

Similar to business cycles, the cyclical ups and downs of REITs can also be observed in many places. When we evaluate whether we should include REITs in our portfolio or not, inclusion of cycle consideration provides extra source of information when we evaluate the value of the investment. Furthermore, as the direct real estate market follows a similar cyclical pattern to indirect properties (REITs) (Wilson and Okunev 1999). One of the property investors' major challenges is the linkage between direct and indirect property investment (Newell and Chau 1996), knowing the correlations between different markets' REITs cycles may also have implications on direct real estate analyses.

With regards to indirect real estate cycles analysis, they include research studies on the interdependence of real estate securities studies in different countries (Liow 2008, Liow 2010). Others shed light on the cyclical relationship between commercial real estate and stocks (Brown and Liow 2001). Nevertheless, research articles on REITs cycles are rare. For example, when we search "REITs cycles" and "real estate investment trusts cycles" in Scholar Google, there is no record on both entries (Scholar Google 2015b, Scholar Google 2015a) apart from the author's recent work (Li and Chow 2015).

11.8 The real estate market in Hong Kong, Japan, the UK and the US and data included for this study

REITs of Hong Kong, Japan, the UK and the US were included in our present study. The US has the longest history, with around 150 publicly traded REITs (Petersen 2004). Previous research suggests that US REITs dividends are sticky. That is mainly because most of the US firms are reluctant to cut dividends and probably because firms are reluctant to cut dividends, firms only increase dividends when there is an increase in earnings. Although these firms may have a target payout ratio, they usually deviate from it and partially adjust dividend payments to higher levels of income over time (Hayunga and Stephens 2009).

In the UK, REITs provide benefits to companies and investors. As UK REITs are listed on the Main Market/AIM, they enjoy all the benefits associated with London's equity markets. REITs, combined with the conventional strengths of London's capital markets nurture the growth of the property investment sector. In the United Kingdom, legislation provided the legal framework for REITs in January 2007 (Baum and Devaney 2008). In the next few years, a number of larger scale listed property companies were converted to UK-REITs and some start-up UK-REITs were created. REITs allow property companies to access equity markets without any tax leakage. Thus, UK-REITs provide real estate investors wider opportunities to access alternative assets. There are a couple of pre-conditions which a company needs to meet in order to become a UK-REIT. These qualifying conditions fall into three categories: balance of business conditions, company conditions and property rental business conditions. Moreover, a potential UK-REIT has to carry out property rental. There is, however, no restriction on location. It can be an overseas property investment business or a UK property investment business. It is required that at least 75% of the group's gross assets must be assets or cash that is involved in property rental; at least 75% of the group's profits must derive from that property's rental income (London Stock Exchange 2015).

In Japan, an amendment to the Investment Trust and Investment Corporation Law in November 2000 offered a new opportunity for Japan's REITs development, making JREITs the first to be introduced on Asian markets (Su *et al.* 2010).

Table 11.1 Descriptive statistics of the REITs indices

REITs index	N	Minimum	Maximum	Mean	Std. deviation
MSCI US	1147	288	964	709	168
Hang Seng	1129	207	776	514	135
Uk	1187	131	524	291	52
Japan	1116	704	1576	966	117

J-REITs are the largest REITs market in Asia and rank fifth largest globally (Newell and Peng 2012).

In Hong Kong, the construction and real estate industry alone account for more than one-fifth of total GDP (Shen and Chung 2006) and the REITs market has become a source of funds for developers. Hong Kong REITs deliver strong risk-adjusted returns and have been robust during the global financial crisis, with superior risk-adjusted returns. Nevertheless, the benefits of diversification were lost during the crisis (Newell *et al.* 2010).

Using daily data from the MSCI US REIT (US), FTSE EPRA/NAREIT UK REITs (UK), Hang Seng REIT (Hong Kong) and J-REIT (Tokyo Stock Exchange) REIT indices (5 days per week), between September 2008 and March 2013 we studied REITs cycles' correlations in three areas of the Bloomberg database. Table 11.1 shows that, since September 2008 UK REITs had the lowest standard deviation after the global subprime financial crisis, while the US recorded the highest. The mean value of REITs was the highest in Japan and the lowest in the UK.

11.8 Research methods and results

Three quantitative methods were used in this research paper: Markov Chain Monte Carlo (MCMC), Hodrick–Prescott filter (HP filter) and Pearson correlation study. The cycle analysis assume that data is continuous, non-existent data points due to holidays or stop trading due to various reasons lead to the missing data points. Hence, MCMC was used to simulate the missing data. Then, an HP filter was used to decompose the time series data into cycle and trend. Finally, Pearson correlation analysis was used to study the correlation between various REITs cycles.

11.8.1 Markov chain Monte Carlo (MCMC)

The research first used Markov chain Monte Carlo (MCMC) to simulate the missing data according to the statistical characteristics of the chained data. The Monte Carlo method is one of the most common ways to solve the complicated mathematical problems by random sampling. It generates pseudo random or random numbers which enter an inverse distribution function. The estimation can be more precise when there is an increased number of iterations, which can be done (Vargas *et al.* 2015, forthcoming). It can perform the needed multidimensional numerical integration to estimate parameters of complex hierarchical models which is nearly impossible to do so by using the simple conventional numerical algorithm (Miaou and Song 2005).

Although previous research suggested that the maximum likelihood estimates (MLE) via the expectation-maximum (EM) gradient algorithms is asymptotically normal and consistent. MCMC is stochastic and iterates between simulating from the conditional distributions of the parameters and latent data. It is considered to be a more ambitious task with higher computation cost than the point estimation that is needed for the EM algorithm approach (Chen *et al.* 2010). According to Yang (2000), there were two steps:

Step 1: The imputation I-step:

The imputation I-step simulated the missing values for each of the observation independently, i.e., if we denoted the missing variables with values of observation *i* by Yi(mis) and variables with observed values by Yi(obs), the I-step drew the values of Yi(mis) from a conditional distribution Yi(mis) based on Yi(obs).

Step 2: The posterior P-step:

The posterior P-step simulated the posterior population covariance matrix and mean vector from the complete sample estimations. The new estimate results were used in the I-step. In absence of the prior information about the parameters, a non-informative prior was then used, e.g. prior information about the covariance matrix stabilized the inference about the mean vector for a near singular covariance matrix (Yang 2000).

The two steps were iterated such that the results were reliable for the imputed data set. The goal was to convert iterates to their stationary distribution and simulated an approximate independent draw of the missing values. With a current parameter estimate β(t) at tth iteration, the I-step draws $Y_{mis}^{(t+1)}$ from p(Ymis|Yobs, $β^{(t)}$) and the P-step draws $β^{(t+1)}$ from p(β|Yobs, $Y_{mis}^{(t+1)}$,). A Markov chain would then be created ($Y_{mis}^{(1)}$, _(1)), ($Y_{mis}^{(2)}$, $β^{(2)}$) and converged in distribution according to p (Y$_{mis}$, β| Y$_{obs}$) (Yang 2000). The results of the maximum, minimum, mean and standard deviation are listed in Table 11.2. All the missing data was filled up, expanding the data to N = 1190.

11.8.2 Hodrick–Prescott (HP) filter

After performing the MCMC procedure, the HP filter was used to decompose the time series data into cycle and trend. As the empirical analysis of business cycle regularities involves a detrending process (García-Ferrer and Río 2001), the present study filtered the real estate time series transaction and GDP data using the HP

Table 11.2 Results of the MCMC analysis

	N	Minimum	Maximum	Mean	Std. deviation
MSCI_US	1190	288	1016	709	168
HS_HK	1190	207	776	514	134
REIT_UK	1190	131	524	291	52
J-REIT	1190	704	1576	965	114

technique. This technique was one of the most common methodologies adopted by business cycle researchers (McGough and Tsolacos 1995) to remove the trend components from the business cycle (Dua 2008). Previous research papers adopted this method to identify business cycles in Japan (Artis and Okuboy 2011) and the US (Partridge and Rickmany 2005). The HP filter estimated the trend in an unobserved components model that has both trend and noise components, with the signal-to-noise ratio "q" set to an arbitrary value. The parameter "q," which was equivalent to the ratio of the variances in the model, was detrended with the associated filter. It is suggested that we should fix q at 1/1600 for quarterly data. The procedure was used to apply the HP filter is the same as that for smoothing a series using a cubic spline (the spline literature focused on methods for choosing a fixed value for the smoothing parameter) (Trimbur 2006).

Given a times series of observations Y_t (t = 1, 2, . . . ,T), the HP filter could be considered to be an additive decomposition $Y_t = Y_t^g + Y_t^c$, where Y_t^g is a trend component and Y_t^c is a cyclical component. The HP filter estimates the trend component via constrained minimization (UCLA Academic Technology Services 2012). According to Gallego and Johnson (2005),

$$\frac{1}{T}\sum_{1=1}^{T}\left(\mathrm{Iny}_t-\mathrm{Iny}_t^*\right)^2-\frac{\partial}{T}\sum_{1=1}^{T}\left[\left(\mathrm{Iny}_{1+1}-\mathrm{Iny}_t^*\right)-\left(\mathrm{Iny}_t-\mathrm{Iny}_{1-1}^*\right)\right]^2 \tag{1}$$

where Y_t^* is the HP filtered time series generated from the actual yt. Its counterpart matrix can be written as

$$y^*-\left(1_t+\partial\left(P'P\right)\right)^{-1}y \tag{2}$$

where P is the TxT matrix represented by a series of concatenated identity and zero matrices, and y and y* are column vectors with dimensions (yx1):

$$P-[1_T|O_{(t-2)x2}]-2[O_{(t-2)x1}|I_{(t-2)}|O_{(t-2)x1}]+[[O_{(t-2)x2}I_{(t-2)}] \tag{3}$$

Results of the HP filter are shown in Table 11.3.

11.8.3 Pearson's correlation analysis

After we obtained the stationary cyclical components, we conducted correlation analysis which was similar to the procedure used in the research study conducted

Table 11.3 Results of HP filter analysis

REITs	N	Minimum	Maximum	Mean	Std. deviation
MSCI_US	1190	−214	196	$3\times10-14$	25
HS_HK	1190	−152	176	$-4\times10-15$	16
REIT_UK	1190	−56	57	$9\times10-15$	9
J-REIT	1190	−337	142	$9\times10-1$	33

by Sayan and Tekin-Koru (2012). This research adopted Pearson's correlation to study the correlation between REITs and GDP cycles. The Pearson product-moment correlation coefficient is an index without dimension and is invariant under the linear transformation of either variable. Pearson's correlation was a mathematical formula developed in 1895 (Rodgers and Nicewander 1988). According to Bolboaca and Jantschi (2006), Pearson's correlation is a measure of the direction and strength of the linear relationship between two variables when one variable is linearly correlated to the other:

$$r_{Prs} = \frac{\sum (X_i - X_m)((Y_i - Y_m))}{[\sum_{i=1}^{n}[(X_i - X_m)^2((Y_i - Y_m)^2]^{1/2}} \tag{4}$$

The Pearson's correlation coefficient varies from -1 to $+1$:

1   $+1$ shows that the variables exhibit a perfect linear relationship;
2   -1 shows that the variables exhibit a perfect linear relationship;
3   0 shows that the variables are not linearly related to one another.

Generally speaking, there is a weak correlation if the correlation coefficient is less than 0.5 and a strong correlation if the correlation coefficient exceeds 0.8.

The Pearson correlation results showed that there were positive and significant cycle relationship between the US, the UK, Hong Kong and Japan REITs. There was strongest relationship between the US and the UK REITs and the weakest relationship between Hong Kong and Japanese REITs.

11.9 Discussion and conclusion

Indirect real estate investment tools such as real estate investment trusts (REITs) provide property investors opportunities to construct internationally invested property portfolios, overcome the shortcomings of direct real estate investment. Nevertheless, studies associated with the effects of international diversification in global REITs investment are relatively new, despite a long history of research in geographical diversification in various types of financial assets (Chang and Chen

Table 11.4 Pearson correlation results of the US, Hong Kong, the UK and Japan REITs cycles

		MSCI_US	HS_HK	REIT_UK	J-REIT
MSCI_US	Pearson correlation	1	.419**	.463**	.247**
HS_HK	Pearson correlation	.419**	1	.202**	.162**
REIT_UK	Pearson correlation	.463**	.202**	1	.310**
J-REIT	Pearson correlation	.247**	.162**	.310**	1

Note: ** Correlation is significant at the 0.01 level (two-tailed), N = 1190 for the four REITs

2014), not to mention the cycles analyses on REITs. Furthermore, from an investor's perspective evaluations and decisions making according to general market movement is vital. Hence, the real estate cycle is an important decision variable for portfolio managers and investors in (Reed and Wu 2010). This chapter used the data in the Bloomberg Database from September 2008 to March 2013 to study the REITs cycles' correlations in the three places. It is the first paper in REITs research which studies the REITs cycles of the first places by filling the missing data by the MCMC method, followed by extracting the cycle via the HP filter. Pearson correlation was used to study the correlations between the REITs cycles in the four places.

The results show that the correlation between REITs cycles in the US, the UK, Hong Kong and Japan were positive and significant: the ups and downs of REITs in each region were similar. When investors estimate the value of including or excluding the REITs in their portfolio, they do not only need to consider the traditional factors such as the returns of REITs but also the REITs cycle in other places. Second, geographical diversification with the aim to lower the risk of investment assets via investment in different locations' REITs market might not be possible. Third, as previous research suggested that there were close relationships between direct and indirect real estate markets, global REITs cycles analyses may be another major concern when surveyors and investors estimate the value of direct real estate investments. Fourth, as suggested by Li (2014), there is a close relationship between housing price indices of the UK and the US, and our research has found that there is a strong correlation between the REITs cycles of the UK and the US. So, we can deduce a certain level of linkage between the US and the UK direct and indirect real estate markets.

Acknowledgement

The authors would like to thank S. Lui and S. Pang for their contribution to this chapter.

11.10 References

Abugri, B. A. and S. Dutta (2014) Are We Overestimating REIT Idiosyncratic Risk? Analysis of Pricing Effects and Persistence. *International Review of Economics and Finance,* 29, 249–259.

Adams, Z., R. Füss and F. Schindler (2015) The Sources of Risk Spillovers among U.S. REITs: Financial Characteristics and Regional Proximity. *Real Estate Economics,* 43, 67–100.

Ahmadi, H., J. A. Kuhle and S. Varsheny (2010) Stress Test for the Financial Optimization Models during the 2009 Recession. *Journal of Business and Economics Research,* 107–114.

Akimov, A., S. Stevenson and M. Zagonov (2015) Public Real Estate and the Term Structure of Interest Rates: A Cross-Country Study. *The Journal of Real Estate Finance and Economics,* 1–38.

Alias, A. and C. Y. Soi Tho (2011) Performance Analysis of REITs: Comparison Between M-REITs and UK-REITs. *Journal of Surveying, Construction and Property,* 2, 38–61.

Allen, M. T., J. Madura and T. M. Springer (2000) REIT Characteristics and the Sensitivity of REIT. *Journal of Real Estate Finance and Economics,* 21, 141–152.

Ambrose, B. W., M. J. Highfield and P. D. Linneman (2005) Real Estate and Economies of Scale: The Case of REITs. *Real Estate Economics,* 33, 323–350.

Ambrose, B. W., D. W. Lee and J. Peek (2007) Comovement After Joining an Index: Spillovers of Nonfundamental Effects. *Real Estate Economics,* 35, 57–90.

Ambrose, B. W. and P. Linneman. (1999). "The Maturing of REITs." In *The Wharton School of the University of Pennsylvania Working Paper Series,* 1–17.

Anderson, R. I., R. Fok, T. Springer and J. Web (2002) Technical Efficiency and Economies of Scale: A Non-parametric Analysis of REIT Operating Efficiency. *European Journal of Operational Research,* 139, 598–612.

Antonakakis, N., I. Chatziantoniou and G. Fillis (2015, forthcoming) Business Cycle Spillovers in the European Union: What is the Message Transmitted to the Core? *The Manchester School.*

Arestis, P. (2015) Is Job Insecurity a Driver of the Housing Cycle? Some Evidence in the Spanish Case. *Panoeconomics* 62, 1–14.

Artis, M. and T. Okuboy (2011) The Intranational Business Cycle in Japan. *Oxford Economic Papers,* 63, 111–133.

Barlevy, G. (2004) The Cost of Business Cycles Under Endogenous Growth. *The American Economic Review,* 94, 964–990.

Baum, A. and S. Devaney (2008) Depreciation, Income Distribution and the UK REIT. *Journal of Property Investment and Finance,* 26, 195–209.

Bers, M. and T. M. Springer (1997) Economies-of-scale for Real Estate Investment Trusts. *Journal of Real Estate Research,* 14, 275–290.

Bolboaca, S.-D. and L. Jantschi (2006) Pearson versus Spearman, Kendall's Tau Correlation Analysis on Structure-Activity Relationships of Biologic Active Compounds. *Leonardo Journal of Sciences,* 9, 179–200.

Brady, P. J. and E. M. Conlin (2004) The Performance of REIT-owned Properties and the Impact of REIT Market Power. *Journal of Real Estate Finance and Economics,* 28, 81–95.

Brown, G. and K. H. Liow (2001) Cyclical Relationship between Commercial Real Estate and Property Stock Prices. *Journal of Property Research,* 18, 309–320.

Burns, A. F. (1969). "The Business Cycle in a Changing World." In *The Nature and Causes of Business Cycles,* ed. A. F. Burns. United States: National Bureau of Economic Research.

Chang, G. D. and C. S. Chen (2014) Evidence of Contagion in Global REITs Investment. *International Review of Economics and Finance,* 31, 148–158.

Chen, H. C., K. Y. Ho, C. L. Lu and C. H. Wu (2005) Real Estate Investment Trusts: An Asset Allocation Perspective. *Journal of Portfolio Management,* 46–54.

Chen, K. and D. Tsang (1988) Interest-Rate Sensitivity of Real Estate Investment Trusts. *Journal of Real Estate Research,* 3, 13–22.

Chen, M. C., C. C. Chang, S. K. Lin and S. D. Shyu (2010) Estimation of Housing Price Jump Risks and Their Impact on the Valuation of Mortgage Insurance Contracts. *The Journal of Risk and Insurance,* 77, 399–422.

Chiang, K. C. H. (2015) What Drives REIT Prices? The Time-Varying Informational Content of Dividend Yields. *Journal of Real Estate Research,* 37, 173–190.

Cotter, J. and R. Roll (2015) A Comparative Anatomy of Residential REITs and Private Real Estate Markets: Returns, Risks and Distributional Characteristics. *Real Estate Economics,* 43, 209–240.

Das, P. K., J. Freybote and G. Marcato (2014) An Investigation into Sentiment-Induced Institutional Trading Behavior and Asset Pricing in the REIT Market. *The Journal of Real Estate Finance and Economics,* 51, 160–189.

Davidoff, T. (2013) Supply Elasticity and the Housing Cycle of the 2000s. *Real Estate Economics,* 41, 793–813.

Devaney, M. (2012) Financial Crisis, REIT Short-sell Restrictions and Event Induced Volatility. *The Quarterly Review of Economics and Finance,* 52: 219–226.

Dua, P. (2008) Analysis of Consumers' Perceptions of Buying Conditions for Houses. *Journal of Real Estate Finance and Economics,* 37, 335–350.

Erol, I. and D. Tirtiroglu (2011) Concentrated Ownership, No Dividend Payout Requirement and Capital Structure of REITs: Evidence from Turkey. *Journal of Real Estate Finance and Economics,* 43, 174–204.

Ewing, B. and J. Payne (2005) The Response of Real Estate Investment Trust Returns to Macroeconomic Shocks. *Journal of Business Research,* 58, 293–300.

Franken, M, J. Z. Bloom and P.D. Erasmus (2011) Factors that Affect South African Real Estate Price Growth Management Dynamics. *Management Dynamics,* 20, 18–32.

Galiniene, B., A. Marcinskas and S. Malevskiene (2006) The Cycles of Real Estate in the Baltic Countries. *Technological and Economic Development of Economy,* 12, 161–167.

Gallego, F. A. and C. A. Johnson (2005) Building Confidence Intervals for Band-Pass and Hodrick–Prescott Filters: An Application Using Bootstrapping. *Applied Economics,* 37, 741–749.

García-Ferrer, A. and A. D. Río (2001) A Cyclical Characterization of Economic Activity in the United States, 1887–1940. *Journal of Economic Studies,* 28, 74–92.

Geltner, D., B. D. MacGregor and G. M. Schwann (2003) Appraisal Smoothing and Price Discovery in Real Estate Markets. *Urban Studies,* 40, 1047–1064.

Genesove, D. and C. Mayer (2001) Loss Aversion and Seller Behavior: Evidence from the Housing Market. *The Quarterly Journal of Economics,* 116, 1233–1260

Ghosh, C., R. Guttery and C. Sirmans (1998) Contagion and REIT Stock Prices. *Journal Of Real Estate Research,* 16, 389–400.

Giliberto, S. M. (1990) Equity Real Estate Investment Trusts and Real Estate Returns. *Journal of Real Estate Research,* 5, 259–264.

Glascock, J., C. Lu and R. So (2000) Further Evidence on the Integration of REIT, Bond, and Stock Returns. *Journal of Real Estate Finance And Economics,* 20, 177–194.

Gordon, J., P. Mosbaugh and T. Canter (1996) Integrating Regional Economic Indicators with the Real Estate Cycle. *The Journal of Real Estate Research,* 12, 469–501.

Hayunga, D. K. and C. P. Stephens (2009) Dividend Behaviour of US Equity REITs. *Journal of Property Research,* 26, 105–123.

Hecq, A., F. C. Palm and J. P. Urbain (2000) Comovements in International Stock Markets: What Can We Learn from a Common Trend-common Cycle Analysis? *De Economist,* 148, 395–406.

Hui, C. M. E. and K. K. K. Chan (2012) Are the Global Real Estate Markets Contagious? *International Journal of Strategic Property Management,* 16, 219–235.

Jirasakuldech, B., R. D. Campbell and J. R. Knight (2006) Are There Rational Speculative Bubbles in REITs? *The Journal of Real Estate Finance and Economics,* 32, 105–127.

Kaiser, R. (1997) The Long Cycle in Real Estate. *Journal of Real Estate Research,* 14, 233–257.

Kamstra, M. J., L. A. Kramer and M. D. Levi (2003) Winter Blues: A SAD Stock Market Cycle. *American Economic Review,* 93, 324–343.

Kim, H., A. S. Mattila and Z. Gu (2002) Performance of Hotel Real Estate Investment Trusts: a Comparative Analysis of Jensen Indexes. *Hospitality Management,* 21, 85–97.

Kodama, M. (2013) How Large Is the Cost of Business Cycles in Developing Countries? *Review of Development Economics*, 17, 49–63.

Lee, Y. H. and T. Y. Pai (2010) REIT Volatility Prediction for Skew-GED Distribution of the GARCH Model. *Expert Systems with Applications*, 37, 4737–4741.

Li, R. Y. M. (2009) The Myth of Fly-by-night Developers in Shanghai and Harbin. *Economic Affairs*, 29, 66–71

Li, R. Y. M. (2010a) *Factors Influencing Developers' Decision to Sell Housing Units with Fittings: Empirical Evidence from China*. Hong Kong: The University of Hong Kong.

Li, R. Y. M. (2010b) A Study on the Impact of Culture, Economic, History and Legal Systems which Affect the Provisions of Fittings by Residential Developers in Boston, Hong Kong and Nanjing. *International Journal of Global Business and Management Research*, 1, 131–141.

Li, R. Y. M. (2012) Econometric Modeling of Risk Adverse Behaviours of Entrepreneurs in the Provision of House Fittings in China. *Australasian Journal of Construction Economics and Building*, 12, 72–82.

Li, R. Y. M. (2014) *Law, Economics and Finance of the Real Estate Market – A Perspective of Hong Kong and Singapore*. Germany: Springer.

Li, R. Y. M. and H. P. Chow (2015) An Economic Analysis on REITs Cycles in Nine Places. *Real Estate Finance*, 32, 23–28.

Li, R. Y. M. and J. Li. (2012). "The Impact of Subprime Financial Crisis on Canada and United States Housing Market and United States Housing Cycle and Economy." In *ICBMG Conference*. Hong Kong.

Li, R. Y. M. and Y. L. Li (2011) An Overview of Regulations That Protect Real Estate Stocks, REITs, Derivatives Investors in Hong Kong. *Lex ET Scientia International Journal*, 18, 200–211.

Ling, D. C. and A. Naranjo (2006) Dedicated REIT Mutual Fund Flows and REIT Performance. *Journal of Real Estate Finance and Economics* 32, 409–433.

Liow, K. H. (2008) Financial Crisis and Asian Real Estate Securities Market Interdependence: Some Additional Evidence. *Journal of Property Research*, 25, 127–155.

Liow, K. H. (2010) Integration among USA, UK, Japanese and Australian Securitised Real Estate Markets: an Empirical Exploration. *Journal of Property Research*, 27, 289–308.

London Stock Exchange. (2015). *REITs*.

Loo, W. K., M. A. Anuar and S. Ramakrishnan (2015) The Dynamic Linkage Among the Asian REITS Market. *Pacific Rim Property Research Journal*, 21, 115–126.

Loutskina, E. and P. E. Strahan (2015) Financial Integration, Housing, and Economic Volatility. *Journal of Financial Economics*, 115, 25–41.

McCue, T. E. and J. L. Kling (1994) Real Estate Returns and the Macroeconomy: Some Empirical Evidence from Real Estate Investment Trust Data, 1972–1991. *Journal Of Real Estate Research*, 9, 277–287.

McGough, T. and S. Tsolacos (1995) Property Cycles in the UK: An Empirical Investigation of the Stylized Facts. *Journal of Property Finance*, 6, 45–62.

Miaou, S. P. and J. J. Song (2005) Bayesian Ranking of Sites for Engineering Safety Improvements: Decision Parameter, Treatability Concept, Statistical Criterion, and Spatial Dependence. *Accident Analysis and Prevention*, 37, 699–720.

Mueller, G. R. (1999) Real Estate Rental Growth Rates at Different Points in the Physical Market Cycle, *Journal of Real Estate Research* 18, 131–150.

Mueller, G. R. (2002) What Will the Next Real Estate Cycle Look Like? *Journal of Real Estate Portfolio Management*, 8 115–125.

Newell, G. and K. W. Chau (1996) Linkages between Direct and Indirect Property Performance in Hong Kong. *Journal of Property Finance*, 7, 9–29.

Newell, G. and H. W. Peng (2012) The Significance and Performance of Japan REITs in a Mixed-Asset Portfolio. *Pacific Rim Property Research Journal,* 18, 21–34.

Newell, G., Y. Wu, K. W. Chau and S. K. Wong (2010) The Development and Performance of REITs in Hong Kong. *Pacific Rim Property Research Journal,* 16, 191–206.

Oppenheimer, P. and T. V. Grissom (1998) Frequency Space Correlation between REITs and Capital Market Indicies. *Journal of Real Estate Research,* 16, 291–309.

Pagliari, J. L. and J. R. Webb (1995) A Fundamental Examination of Securitized and Unsecuritized Real Estate. *Journal of Real Estate Research,* 10, 381–426.

Partridge, M. D. and D. S. Rickmany (2005) Regional Cyclical Asymmetries in An Optimal Currency Area: An Analysis Using US State Data. *Oxford Economic Papers,* 57, 373–397.

Petersen, A. (2004) The Major Issues Facing the Successful Introduction of the UK REIT. *Briefings in Real Estate Finance* 4, 8–20.

Reed, R. and H. Wu (2010) Understanding Property Cycles in a Residential Market. *Property Management,* 28, 33–46.

Renaud, B. (1997) The 1985 to 1994 Global Real Estate Cycle: An Overview. *Journal of Real Estate Literature,* 5, 13–44.

Reynolds, T. (1997) REITs Under a Brighter Light. *Journal of Financial Planning,* 10, 68–77.

Rodgers, J. L. and W. A. Nicewander (1988) Thirteen Ways to Look at the Correlation Coefficient. *The American Statistician,* 42, 59–66.

Sayan, S. and A. Tekin-Koru (2012) Remittances, Business Cycles and Poverty: The Recent Turkish Experience. *International Migration* 50, e151–e176.

Schindler, F. (2009) Correlation Structure of Real Estate Markets Over Time. *Journal of Property Investment and Finance,* 27, 579–592.

Scholar Google. (2015a). "Real Estate Investment Trusts Cycle".

Scholar Google. (2015b). "REITs Cycle".

Scott, J. L., R. I. Anderson and J. R. Webb (2005) The Labor–Leisure Choice in Executive Compensation Plans: Does Too Much Pay Reduce REIT Performance? *Journal of Economics and Business,* 57, 151–163.

Shen, G. Q. P. and J. K. H. Chung (2006) A Critical Investigation of the Briefing Process in Hong Kong's Construction Industry. *Facilities,* 24, 510–522.

Stevenson, S. (2001) The Long-term Advantage to Incorporating Indirect Securities in Direct Real Estate Portfolios. *Journal of Real Estate Portfolio Management,* 7, 5–16.

Su, H. M., C. M. Huang and T. Y. Pai (2010) The Hybrid Characteristic of REIT Returns: Evidence from Japanese and U.S. States Markets. *Journal of Real Estate Literature,* 18, 77–98.

Sun, L., S. D. Titman and G. J. Twite (2015) REIT and Commercial Real Estate Returns: A Postmortem of the Financial Crisis. *Real Estate Economics,* 43, 8–36.

Tang, C. H. H. and S. C. S. Jang (2008) The Profitability Impact of REIT Requirements: A Comparative Analysis of Hotel REITS and Hotel C-Corporations. *International Journal of Hospitality Management,* 27, 614–622.

Tobias, B., F. Meik and V. B. Eduardo (2009) REITs and the Financial Crisis: Empirical Evidence from the U.S. *International Journal of Business and Management,* 4, 3–10.

Trimbur, T. M. (2006) Detrending Economic Time Series: A Bayesian Generalization of the Hodrick–Prescott Filter. *Journal of Forecasting,* 25, 247–273.

Tsai, H. J., C. J. Lin and M. C. Chen (2012) Contagion between Stock and REITs Markets During the Subprime and Financial Crisis. *International Research Journal of Finance and Economics,* 102, 82–103.

UCLA Academic Technology Services. (2012). What does Cronbach's Alpha Mean?

Vargas, J. P., J. C. Koppe, S. Pérez and J. P. Hurtado (2015, forthcoming) Planning Tunnel Construction Using Markov Chain Monte Carlo (MCMC). *Mathematical Problems in Engineering.*

Wernecke, M., N. Rottke and C. Holzmann (2004) Incorporating the Real Estate Cycle into Management Decisions-Evidence from Germany. *Journal of Real Estate Portfolio Management,* 10, 171–186.

Wilson, P. and J. Okunev (1999) Spectral Analysis of Real Estate and Financial Assets Markets. *Journal of Property Investment and Finance,* 17, 61–74.

Wu, P. S., C. M. Huang and C. L. Chiu (2011) Effects of Structural Changes on the Risk Characteristics of REIT Returns. *International Review of Economics and Finance,* 20, 645–653.

Yong, J. and A. Singh (2015) Interest Rate Risk of Australian REITS: A Panel Analysis. *Pacific Rim Property Research Journal,* 21, 77–88.

Zhu, H. (2002) The Case of the Missing Commercial Real Estate Cycle. *BIS Quarterly Review,* September, 56–66.

12 Conclusion

Should mainstream economists neglect and undermine real estate economics? The 'Wh' questions in the international housing markets, macroeconomy and econometric models context

Rita Yi Man Li and Kwong Wing Chau

History confirms that property markets are fundamentally, unavoidably and basically volatile. Booms and busts in international housing markets have occurred as a result of railway developments, economic and financial depressions, wars, gold rushes, oil crises, interest rates, investment waves, inflation and taxation (Bloch 1997). In recent years, globalization has affected cities in various different ways through the internationalization of labor markets, capital flows, information, raw materials, commodity markets and organizational structures management (McGreal *et al.* 2002). There are many aspects, such as technological capability, a highly trained workforce, and internationalized housing (Abdul-Aziz and Awil 2010).

In view of these, several schools of thoughts have been developed to examine, investigate and provide vivid explanations for such phenomena: for example, Dunning's eclectic paradigm, network theory, stage growth theory, and diamond framework. The eclectic paradigm suggests that some locations have locational merits, which attract multinational companies to invest. Hence, many of the real estate developers ventured overseas to meet the untapped housing demand, as homebuyers not only buy local housing units but also dwellings in the overseas markets (Abdul-Aziz and Awil 2010).

Modern global property investors have made an important impact on city development. Their attitudes affect global capital mobility as property investment is embodied in foreign direct investment. Changes in the flows of funds usually facilitate the expansion of this kind of investment. On top of that, shifts in the geographic structure of the global economy have created more opportunities for property investment, in particular in the Asia-Pacific region. Pressed by global competition, national governments have deregulated and liberalized the policies on property investments. These policy changes have increased opportunities for global housing investment (Kim *et al.* 2015b).

Previous research indicated that residential property risk profiles could be lowered considerably with a well-diversified residential property portfolio, according to an efficient frontier model (Higgins and Fang 2012). Hence, international property investment has become more important in recent years. Global property investors – which include Jones Lane LaSalle, RREEF Property Trusts, Internationale Nederlanden Groep – have expanded from traditional mature property markets – such as Europe, the US, the UK and Australia – to emerging property markets, particularly in Asia, leading to significant economic growth and an increase in market maturity. Asian property, for example, accounts for close to one-fifth of global investible property and Asia accounts for over one-fourth of the total number of global commercial property transactions in 2008. While the Asian stock markets account for approximately one-third of the global stock market capitalization in 2008, Asian property securities markets account for nearly half of the global property securities. International property diversification is more effective in Asian property markets than the traditional ones. Furthermore, diversification opportunities via investments in several Asian countries' property securities open the door to global real estate investments. It has been further enhanced by an increase in property market maturity and the introduction of REITs in Asian countries (Newell *et al.* 2009).

In Seoul, many housing units were bought by foreigners from the US, Canada, Taiwan, China, Australia, Japan and New Zealand. In the UK, many landmark buildings are owned by overseas investors. Many of those investors come from the US, Japan, China, Germany, Sweden, the Netherlands and the Middle East (McAllister 2000). In Karachi, migration has continued in the last decade for a variety of reasons, such as anti-Taliban army action and issues of safety (Hasan 2015). In summary, there is an obvious increase in cross-border property investment. The relative attractiveness of indirect/direct international real estate investment depends on the preferences and objectives of individuals or investing institutions (McAllister 2000).

International investment motivation may be considered in four dimensions. First, international investment plays an important role in diversification. New technology and financial deregulation offers investors golden opportunities for geographical investment diversification. It provides the opportunity to improve the risk-adjusted returns. Beyond doubt, it is one of the major goals for international portfolio investment in the modern real estate sector (McAllister 2000).

Furthermore, investors may enjoy higher income yields or returns than investing in the domestic market. Many investors consider the opportunity to obtain high returns as an important factor in engaging in international real estate investment. Third, some of the investors may be motivated to invest in international markets as one of the operational requirements in core business. As the firms expand, acquiring liabilities in different markets may enhance a companies' growth (McAllister 2000).

Fourth, foreign housing investment can be understood as the outcome of inter-related activities: international migration has increased significantly due to new divisions of labor and technological breakthrough. Global movements in the labor market have been accelerated since the 1980s between the first- and the third-world due to the increase in income gaps between cities around the world, large-scale immigration takes place. High income cities absorb immigrants and international migration supplements the labor shortage in some of the countries which suffer from ageing populations (Kim *et al.* 2015a).

As the flowering and withering of the housing markets have an indispensable, inevitable crucial relationship with the macroeconomy, analyzing the existing housing market price patterns provides valuable insights to study the various housing attributes which may be more attractive to consumers (Fullerton and Villalobos 2011). This book summarized the impact of various macroeconomic data on the housing prices and the application of econometric techniques. Chapter 2 shed light on the application of various econometric models to the analysis of the housing market. In this chapter, we attempted to draw a simple conclusion for the whole book according to the 'Wh' questions: who, where, why and what.

Who are the major players that affect the flourishing and withering of the housing market? *Who* should study the relationship between the macro economy and the international housing market? In spite of the fact that many of these chapters' data did not come from the same country, we concluded that macroeconomic data did play an important role in the rise and fall of housing prices. Hence, when we study housing markets in a global context, strong economic background does help real estate practitioners in understanding the changes in housing markets. Likewise, economists should also grasp a basic understanding of housing markets as they are an important part of our economy. As one of the chapters noted, the "housing cycle *is* the business cycle," whilst we may not put an equal sign between the housing market and real gross domestic product, none of us should dismiss the close relationship between them. Indeed, while many of the so-called mainstream economists undermine or neglect this relationship, should economists study the real estate market? The answer should be affirmative after reading this book.

How to study the relationships between them? Econometric modeling plays an important role in scientific analysis. For example, Chapter 1 suggested that there are a variety of econometrics methods in studying the relationship between them. Chapter 4 utilized the State Space model to perform the property stock forecast. Chapter 10 illustrated the use of the HP filter to filter the cycle out from time series data. Pearson correlation analysis was adopted to study the relationship between the REITs cycles.

The major reasons of *why* we have to use econometric methods are illustrated by our book chapters. In absence of the filter, we cannot observe the relationships of various REITs cycles in different countries. For example, Chapter 10 used the HP filter to extract the cycles from time series data. Furthermore, "change" always occurs in society. Previous research argued that the major changes in economic policy led to a more market driven demand for housing investment in Sweden at the end of the 1980s and the early 1990s (Lennart 2006).

Chapter 7 adopted the Chow test and Quantile regression to study the impact of the financial crisis on the housing market of Germany and Norway. As we do not have a crystal ball to make forecasts and we often fall short of adequate market data or can only find some data with low frequency (Adams and Füss 2010), it is of both academic and practical interest to explore possible methods of performing such forecasting. In Chapter 3, we illustrated the use of the State Space model in forecasting property stock prices when some data such as GDP and listed real estate company data is less frequent than other data, e.g. daily stock prices.

Chapter 9 utilized the Cobb–Douglas Model to study the input–output relationship with regards to the various factors (input), which drives housing prices (output) up and down. Likewise, whilst we may have a gut feeling on why some of the housing entrepreneurs provide fittings but others do not, the analysis may not be systematic. Chapter 2 illustrated the importance of the Probit model in analyzing the factors which lead to the creation of bare units.

The next question when studying the relationship between the housing market and the macroeconomy is **where** should our focus be? Does it mean that we only need to pay attention to local issues? It is inevitable and unavoidable that local issues play a very important role in housing demand as regulation and other local issues may affect housing demand and supply (Li 2014). For example, Chapter 5 illustrated the impact of local news on property prices and Chapter 6 showcased the importance of the railway on raising the housing prices. Chapter 7 illustrated that a spectacular view pushed up housing prices despite the existence of air and noise pollutants from a nearby main road.

What are the major macroeconomic factors that affect the housing market? Previous research suggested that proxies for housing demand included market and population growth, purchasing power, size of market and gross domestic growth (Abdul-Aziz and Awil 2010). And several demand factors also affect the housing prices. Chapters 4 and 5 revealed that local news and culture often affect housing prices. Chapter 9 showed the shock of interest rates affecting housing prices in the Czech Republic and South Africa.

Apart from that, as many of the financial markets are open, global financial investors can invest and speculate in stock markets worldwide. The contagion effect has become one of the most important international issues. For example, the Asian Financial crisis started in Thailand but the shocks spread to Hong Kong and many other Asian countries. This book has shown that the spillover effect of the financial crisis in the US adversely affected Norway and Germany, falling in line with the author's previous research in New Zealand and Canada (Li and Li 2012, Li 2012).

12.1 References

Abdul-Aziz, A. R. and A.-U. Awil (2010) Locational Considerations and International Malaysian Housing Developers. *Journal of Financial Management of Property and Construction,* 15, 7–20.

Adams, Z. and R. Füss (2010) Macroeconomic Determinants of International Housing Markets. *Journal of Housing Economics,* 19, 38–50.

Bloch, B. (1997) Volatility in the Residential Housing Market: An International Perspective. *Property Management,* 15, 12–24.

Fullerton, T. M. and E. Villalobos (2011) Street Widths, International Ports of Entry and Border Region Housing Values. *Journal of Economic Issues,* 45, 493–509.

Hasan, A. (2015) Land Contestation in Karachi and the Impact on Housing and Urban Development. *Environment and Urbanization,* 27, 217–230.

Higgins, D. and F. Fang (2012) Analysing the Risk and Return Profile of Chinese Residential Property Markets. *Pacific Rim Property Research Journal,* 18, 149–162.

Kim, H. M., S. S. Han and K. B. O' Connor (2015a) Foreign Housing Investment in Seoul: Origin of Investors and Location of Investment. *Cities,* 42, 212–223.

Kim, H. M., K. B. O'Connor and S. S. Han (2015b) The Spatial Characteristics of Global Property Investment in Seoul: A Case Study of the Office Market. *Progress in Planning,* 97, 1–42.

Lennart, B. B. (2006) The Q Theory and the Swedish Housing Market – An Empirical Test. *Journal of Real Estate Finance and Economics,* 33, 329–344.

Li, R. Y. M. (2012). "Chow Test Analysis on Structural Change in New Zealand Housing Price During Global Subprime Financial Crisis." In *18th Annual Pacific Rim Real Estate Society Conference.* Adelaide.

Li, R. Y. M. (2014). *Law, Economics and Finance of the Real Estate Market – A Perspective of Hong Kong and Singapore.* Germany: Springer.

Li, R. Y. M. and J. Li. (2012). "The Impact of Subprime Financial Crisis on Canada and United States Housing Market and United States Housing Cycle and Economy." In *ICBMG Conference.* Hong Kong.

McAllister, P. (2000) Is Direct Investment in International Property Markets Justifiable? *Property Management,* 18, 25–33.

McGreal, S., A. Parsa and R. Keivani (2002) Evolution of Property Investment Markets in Central Europe: Opportunities and Constraints. *Journal of Property Research,* 19, 213–230.

Newell, G., K. W. Chau, S. K. Wong and K. H. Liow (2009) The Significance and Performance of Property Securities Markets in the Asian IFCs. *Journal of Property Research,* 26, 125–148.

Index